Probably Not

Probably Not

Future Prediction
Using Probability and
Statistical Inference

Lawrence N. Dworsky

Second Edition

Registered Office

John Wiley & Sons, Inc., 111 River Street, Hoboken, NJ 07030, USA

Editorial Office

111 River Street, Hoboken, NJ 07030, USA

For details of our global editorial offices, customer services, and more information about Wiley products visit us at www.wiley.com.

Wiley also publishes its books in a variety of electronic formats and by print-on-demand. Some content that appears in standard print versions of this book may not be available in other formats.

Library of Congress Cataloging-in-Publication Data

Names: Dworsky, Lawrence N., 1943– author.
Title: Probably not : future prediction using probability and statistical inference / Lawrence N. Dworsky.
Description: Second edition. | Hoboken, NJ : Wiley, 2019. | Includes index. |
Identifiers: LCCN 2019019378 (print) | LCCN 2019021721 (ebook) | ISBN 9781119518136 (Adobe PDF) | ISBN 9781119518129 (ePub) | ISBN 9781119518105 (pbk.)
Subjects: LCSH: Prediction theory. | Probabilities–Problems, exercises, etc. | Mathematical statistics–Problems, exercises, etc.
Classification: LCC QA279.2 (ebook) | LCC QA279.2 .D96 2020 (print) | DDC 519.2/87–dc23
LC record available at https://lccn.loc.gov/2019019378

To all the people who realize, or want to learn, that Probability and Statistics isn't just a mathematics discipline, it's a way to see and better understand the world.

Contents

Acknowledgments

Ten years ago my wife, Suzanna, patiently encouraged me to write this book during the many months it took me to actually get started. She listened to my ideas, read and commented on drafts and perhaps most importantly, suggested the title.

This year she once again supplied the encouragement and support for me to repeat the process.

The folks at John Wiley were very encouraging and helpful at the start and continue to be so.

I am grateful to all of you.

—Lawrence N. Dworsky

About the Companion Website

This book is accompanied by a companion website:

www.wiley.com/go/probablynot2e

The website includes the following materials for instructors:

- Solutions for all the problems in the chapters.

Introduction

I have always been interested in how well we know what we say we know, how we acquire data about things we want to know, and how we react to and use this data. I was surprised when I first realized that many people are not only uninterested in these issues, but are actually averse to learning about them. I have since realized that we seem to be genetically programmed on the one hand to intelligently learn how to acquire data and use it to our advantage while on the other hand to stubbornly refuse to believe what some simple calculations and/or observations tell us.

My first conclusion is supported by humanity's march through history, learning about agriculture and all the various forms of engineering and biology and using this knowledge to make life better and easier. My latter conclusion comes from seeing people sitting on stools in front of slot machines at gambling casinos, many of whom are there that day because the astrology page in the newspaper told them that this was "their day."

This book is mostly about probability, with just a few chapters dedicated to an introduction to the vast field of statistical inference.

Many excellent books on probability and statistics are available today. These books fall into two general categories. One category is textbooks. Textbooks are heavily mathematical with derivations, proofs and problem sets, and an agenda to get you through a term's course work. This is just what you need if you are taking a course.

The other category is books that are meant for a more casual audience – an audience that's interested in the topic but isn't interested enough to take up formal study. We're told that people today have "mathephobia," and the books that appeal to these people try very hard to talk around the mathematics without actually presenting any of it. Probability and statistics are mathematical topics. A book about these subjects without math is sort of like a book about French grammar without any French words in it. It's not impossible, but it sure is doing things the hard way.

Probably Not: Future Prediction Using Probability and Statistical Inference,
Second Edition. Lawrence N. Dworsky.
© 2019 John Wiley & Sons, Inc. Published 2019 by John Wiley & Sons, Inc.
Companion website: www.wiley.com/go/probablynot2e

I try to split the difference between the above two choices. This is a textbook that includes more than the typical textbook does about real-world topics. There is some math involved. How much? Introductory high school algebra along with a little patience in learning some new notation should comfortably get you through it. Even if you don't remember how to do things yourself, just realizing what I'm doing and accepting that I'm probably doing it right should be enough. You should recognize a square root sign and sort of remember what it means. You don't need to know how to calculate a square root – these days everybody does it on a pocket calculator or a computer spreadsheet anyway. You should be able to read a graph.

In a very few cases some elementary calculus is needed to get from point A to point B. For readers with no calculus background, I have included a short Appendix that introduces a couple of necessary concepts and notations. This certainly is not an alternative to learning calculus, but in this limited context it is enough to let the reader understand what is being calculated and why.

As I discuss in the first chapter, probability is all about patterns of things such as what happens when I roll a pair of dice many times, or what the life expectancies of the population of the United States looks like, or how long to expect to wait for your bus. Just as a course in music with some discussions of rhythm and harmony helps you "feel" the beauty of the music, a little insight into the mathematics of the patterns of things in our life can help you feel the beauty of these patterns as well as to plan things that are unpredictable (when will the next bus come along and how long will I have to stand in the rain to meet it?) as best possible.

Probabilistic considerations show up in several areas of our lives. Some are explicitly from nature, such as daily rainfall or life expectancy. Some are from human activities, everything from gambling games to manufacturing tolerances. Some are from nature but we don't see them until we "look behind the green curtain." This includes properties of gases (e.g. the air around us) and the basic atomic and subatomic nature of matter.

Probability and statistics often involves examining sets of data and summarizing in order to draw a conclusion. For example "the average daily temperature in Coyoteville is 75 degrees Fahrenheit." Coyoteville sounds like a great place to live until you learn that the temperature during the day peaks at 115 degrees while at night it drops to 35 degrees. In some cases we will be adding insight by summarizing a data set, in some cases we will be losing insight; we have to study the summarizing processes to understand this.

The field of probability and statistics has a very bad reputation ("Lies, Damned Lies, and Statistics"[1]). It is easy to manipulate conclusions by simply omitting

---

1 This quote is usually attributed to Benjamin Disraeli, but there seems to be some uncertainty here. I guess that – considering the book you're now holding – I should say that "There is a high probability that this quote should be attributed to Benjamin Disraeli."

some of the data, or to perform the wrong calculations correctly, or to misstate the results – any and all of these actions are possibly innocent – because some problems are complicated and subtle. The materials to follow show what information is needed to draw a conclusion and what conclusion(s) can and cannot be drawn from certain information. Also, I'll show how to reasonably expect that sometimes, sometimes even inevitably, as the bumper stickers say, *stuff* happens.

I spend a lot of time on simple gambling games because, even if you're not a gambler, there's a lot to be learned from the simplest of random events, the results of coin flips.

I've chosen some examples you don't usually see in probability books. I look at waiting for a bus, life insurance, scheduling appointments, etc. I want to convey that we live in a world where many of our daily activities involve random processes and the statistics involved with them.

Finally, I introduce some topics that show how much of our physical world is based on the consequences of random processes. These topics include gas pressure, heat engines, and radioactive decay. Although they're not relevant to things you might actually do such as meeting a friend for lunch or waiting in a doctor's office, I hope you'll find reading about them to be interesting.

Virtually every mathematics text has problem sets at the ends of chapters. Probability and statistics is an unusual discipline in that there are many people who love being challenged by problems in these areas. Many books on the market and many websites offer everything from beginners' to experts' problems. Consequently, this book does not have an exhaustive number of problems. The problems it has were chosen to make a point; the solutions are almost always much more complete than "the answer is 6.2." Many solutions contain discussions, graphs, and comments. The problems and solutions are more like extensions of the chapters than drills.

Most of the problems are designed to be solvable with a paper and pencil. In some cases there are just too many numbers involved to do this, and a spreadsheet becomes the preferred tool. The spreadsheet that is part of any major computer office suite will be adequate. In addition to dealing with hundreds of numbers quickly and easily, modern spreadsheets have built-in capabilities for looking up logarithms, binomial probabilities, etc. Some comfort in working with a spreadsheet is an indispensable requirement. In a few cases, tens of thousands of numbers are involved in a problem. Here, a programming language is the best way to go. There aren't many of these problems, but the few that there are must be passed over if you don't have some programming ability.

This second edition differs from the first edition in four aspects:

1) The text has been thoroughly edited.
2) Several chapters have been eliminated because upon rereading them a decade after writing them, I decided they were tedious and the points being made just aren't worth the number of pages involved.

3) Several new topics have been added, including two new chapters on Bayesian statistics; this probably is the most significant change between editions. Also, the Benford's Law section has been significantly rewritten.

4) Problem and solution sets have been added. A few chapters do not have problem and solution sets. These chapters are meant as discussions of relevant real-world examples of probability and random variable happenings.

One last comment: There are hundreds – if not thousands – of clever probability problems *out there*. I've included several popular ones (including the shared birthday, the prize behind one of 3 doors, the lighthouse location) and discuss how to solve them. When first confronted with one of these problems, I inevitably get it wrong. In my own defense, when I sit down and work things out carefully, I (usually) get it right. This is a tricky subject. Maybe that's why I find it so interesting.

1

An Introduction to Probability

Predicting the Future

The term Predicting the Future conjures up images of veiled women staring into hazy crystal balls or bearded men with darting eyes passing their hands over cups of tea leaves or something else equally humorously mysterious. We call these people fortune tellers, and relegate their professions to the regime of carnival side-show entertainment, along with snake charmers and the like. For party entertainment, we bring out a Ouija board; everyone sits around the board in a circle and watches the board extract its mysterious energy from our hands while it answers questions about things-to-come.

On the one hand, we all seem to have firm ideas about the future based on consistent patterns of events that we have observed. We are pretty sure that there will be a tomorrow, and that our clocks will all run at the same rate tomorrow as they did today. If we look in the newspaper (or these days, on the Internet), we can find out what time the sun will rise and set tomorrow – and it would be difficult to find someone willing to place a bet that this information is not accurate. On the other hand, whether or not you will meet the love of your life tomorrow is not something you expect to see accurately predicted in the newspaper.

We seem willing to classify predictions of future events into categories of the knowable and the unknowable. The latter category is left to carnival fortune tellers to illuminate. The former category includes predictions of when you'll next need a haircut, how much weight you'll gain if you keep eating so much pizza, etc.

But, there does seem to be an intermediate area of knowledge of the future. Nobody knows for certain when you're going to die. An insurance company, however, seems able to consult its mystical Actuarial Tables and decide how much to charge you for a life insurance policy. How can an insurance company

Probably Not: Future Prediction Using Probability and Statistical Inference,
Second Edition. Lawrence N. Dworsky.
© 2019 John Wiley & Sons, Inc. Published 2019 by John Wiley & Sons, Inc.
Companion website: www.wiley.com/go/probablynot2e

do this if nobody knows when you're going to die? The answer seems to lie in the fact that if you study thousands of people similar in age, health, life style, etc., you can calculate an average life span – and that if the insurance company sells enough insurance policies with rates based upon this average, in a financial sense this is as good as if the insurance company knows exactly when you are going to die. There is, therefore, a way to describe life expectancies in terms of the expected behavior of large groups of people of similar circumstances.

When predicting future events, you often find yourself in these situations. You know something about future trends but you do not know exactly what is going to happen. If you flip a coin, you know you'll get either heads or tails, but you don't know which. If you flip 100 coins, or equivalently flip one coin 100 times, however, you'd expect to get approximately 50 heads and 50 tails.

If you roll a pair of dice[1] you know that you'll get some number between two and twelve, but you don't know which number you'll get. You do know that it's more likely that you'll get six than that you'll get two.

When you buy a new light bulb, you may see written on the package "estimated lifetime 10,000 hours." You know that this light bulb might last 10 346 hours, 11 211 hours, 9587 hours, 12 094 hours, or any other number of hours. If the bulb turns out to last 11 434 hours you won't be surprised, but if it only lasts 1000 hours you'd probably switch to a different brand of light bulbs.

There is a hint in each of these examples which shows that even though you couldn't accurately predict the future, you could find some kind of pattern that teaches you something about the nature of the future. Finding these patterns, working with them, and learning what knowledge can and cannot be inferred from them is the subject matter of the study of probability and statistics.

We can separate our study into two classes of problems. The first of these classes is understanding the likelihood that something might occur. We'll need a rigorous definition of likelihood so that we can be consistent in our evaluations. With this definition in hand, we can look at problems such as "How likely is it that you can make money in a simple coin flipping game?" or "How likely is it that a certain medicine will do you more good than harm in alleviating some specific ailment?" We'll have to define and discuss random events and the patterns that these events fall into, called Probability Distribution Functions (PDFs). This study is the study of Probability.

The second class of problems involves understanding how well you really know something. We will only discuss quantifiable issues, not "does she really love me?" or "is this sculpture a fine work of art?"

The uncertainties in how well we know something can come from various sources. Let's return to the example of light bulbs. Suppose you're the manufacturer of these light bulbs. Due to variations in materials and manufacturing

---

1 In common practice, the result of rolling a pair of dice is the sum of the number of dots facing up when the dice come to rest.

processes, no two light bulbs are identical. There are variations in the lifetime of your product that you need to understand. The easiest way to study the variations in lifetime would be to run all your light bulbs until they burn out and then look at the data, but for obvious reasons this is not a good idea. If you could find the pattern by just burning out some (hopefully a small percentage) of the light bulbs, then you have the information you need both to truthfully advertise your product and to work on improving your manufacturing process.

Learning how to do this is the study of Statistics. We will assume that we are dealing with a stationary random process. In a stationary random process, if nothing causal changes, we can expect that the nature of the pattern of the data already in hand will be the same as the nature of the pattern of future events of this same situation, and we use statistical inference to predict the future. In the practical terms of the example of our light bulb manufacturer, we are saying that as long as we don't change anything, the factory will turn out bulbs with the same distribution of lifetimes next week as it did last week. This assertion is one of the most important characteristics of animal intelligence, namely the ability to discern and predict based upon patterns.

The light bulb problem also exemplifies another issue that we will want to examine. We want to know how long the light bulb we're about to buy will last. We know that no two light bulbs are identical. We also realize that our knowledge is limited by the fact that we haven't measured every light bulb made. We must learn to quantify how much of our ignorance comes from each of these factors and develop ways to express both our knowledge and our lack of knowledge.

Rule Making

As the human species evolved, we took command of our environment because of our ability to learn. We learn from experience. Learning from experience is the art/science of recognizing patterns and then generalizing these patterns into a rule. In other words, the pattern is the relevant raw data that we've collected. A rule is what we create from our analysis of the pattern that we then use to predict the future. Part of the rule are (or some) preferred extrapolations and responses. Successful pattern recognition is, for example, seeing that seeds from certain plants, when planted at the right time of the year and given the right amount of water, will yield food; and that the seed from a given plant will always yield that same food. Dark, ominous looking, clouds usually precede a fierce storm and it's prudent to take cover when such clouds are seen. Also, leaves turning color and falling off the trees means that winter is coming and preparations must be made so as to survive until the following spring.

If we noticed that every time it doesn't rain for more than a week our vegetable plants die, we would generate a rule that if there is no rain for a week, we need to irrigate or otherwise somehow water the vegetable garden. Implicit in this is that somewhere a hypothesis or model is created. In this case our model is that plants need regular watering. When the data is fit to this model, we quantify the case that vegetable plants need water at least once a week, and then the appropriate watering rule may be created.

An interesting conjecture is that much, if not all, of what we call the arts came about because our brains are so interested in seeing patterns that we take delight and often find beauty in well-designed original patterns. Our eyes look at paintings and sculptures, our ears listen to music, our brains process the language constructs of poetry and prose, etc. In every case we are finding pleasure in studying patterns. Sometimes the patterns are clear, as in a Bach fugue. Sometimes the patterns are harder to recognize, as in a surrealistic Picasso painting. Sometimes we are playing a game looking for patterns that may or may not be there – as in a Pollack painting. Perhaps this way of looking at things is sheer nonsense, but then how can you explain why a good book or a good symphony (or rap song if that's your style) or a good painting can grab your attention, and in some sense, please you? The arts don't seem to be necessary for our basic survival; why do we have them at all?

A subtle rustling in the brush near the water hole at dusk sometimes – but not always – means that a man-eating tiger is stalking you. It would be to your advantage to make a decision and take action. Even if you're not certain that there's really a tiger present, you should err on the cautious side and beat a hasty retreat; you won't get a second chance. This survival skill is a good example of our evolutionary tendency to look for patterns and to react as if these patterns are there, even when we are not really sure that they indeed are there. In formal terms, you don't have a lot of data, but you do have anecdotal information.

Our prehistoric ancestors lived a very provincial existence. Life spans were short; most people did not live more than about 30 years. They didn't get to see more than about 10 000 sunrises. People outside their own tribe (and possibly some nearby tribes) were hardly ever encountered, so that the average person never saw more than a few hundred people over the course of a lifetime. Also, very few people (other than members of nomadic tribes) ever traveled more than about 50 miles from where they were born. There are clearly many more items that could be added to this list, but the point has probably been adequately made: Peoples' brains never needed to cope with situations where there were hundreds of thousands or millions of data points to reconcile.

However, in today's world things are very different: A state lottery could sell a hundred million tickets every few months. There are about 7 billion (that's seven thousand million) people on the earth. Many of us (at least in North America and Western Europe) have traveled thousands of miles from the place of our birth many times; even more of us have seen movies and TV shows that depict places and people all over the world. Due to the ease with which people move around, a disease epidemic is no longer a local issue. Also, because we are aware of the lives of so many people in so many places, we know about diseases that attack only one person in a hundred thousand and tragedies that occur just about anywhere. If there's a vicious murderer killing teenage girls in Boston, then parents in California, Saskatoon, and London hear about it on the evening news and worry about the safety of their daughters.

When dealing with unlikely events spread over large numbers of opportunities, your intuition can and does often lead you astray. Since you cannot easily comprehend millions of occurrences, or lack of occurrences, of some event, you tend to see patterns from a small numbers of examples; again the anecdotal approach. Even when patterns don't exist, you tend to invent them; you are using your "better safe than sorry" prehistoric evolved response. This could lead to the inability to correctly make many important decisions in your life: What medicines or treatments stand the best chance of curing your ailments? Which proffered medicines have been correctly shown to be useful, which are simply quackery? Which environmental concerns are potentially real and which are simple coincidence? Which environmental concerns are no doubt real but probably so insignificant that we can reasonably ignore them? Are sure bets on investments or gambling choices really worth anything? We need an organized methodology for examining a situation and coping with information, correctly extracting the pattern and the likelihood of an event happening or not happening to us, and also correctly processing a large set of data and concluding, when appropriate, whether or not a pattern is really present.

We want to understand how to cope with a barrage of information. We need a way of measuring how sure we are of what we know, and when or if what we know is adequate to make some predictions about what's to come.

Random Events and Probability

This is a good place to introduce the concepts of random events, random variables, and probability. These concepts will be wrung out in detail in later chapters, so for now let's just consider some casual definitions.

An *event* is a *particular occurrence* of some sort out of a larger set of *possible occurrences*. Some examples are:

- Will it rain tomorrow? The full set of possible occurrences is the two events Yes – it will rain, and No – it won't rain.
- When you flip a coin, there are two possible events. The coin will either land head side up or tail side up (typically referred to as "heads" or "tails").

- When you roll one die, then there are six possible events, namely the six faces of the die that can land face up: i.e. the numbers 1, 2, 3, 4, 5, and 6.
- When you play a quiz game where you must blindly choose "door A, door B, or door C" and there is a prize hiding behind only one of these doors, then there are 3 possible events: The prize is behind door A, it's behind door B, or it's behind door C.

Variable is a name for a number that can be assigned to an event. If the events themselves are numbers, e.g. the six faces of the die mentioned above, then the most reasonable thing to do is to simply assign the variable numbers to the event numbers. A variable representing the days of the year can take on values 1, 2, 3, ... all the way up to 365. Both of these examples are of variables that must be integers, i.e. 4.56 is not an allowed value for either of them.

There are of course cases where a variable can take on any value, including fractional values, over some range; for example, the possible amount of rain that fell in Chicago last week can be anything from 0 to 15 in. (I don't know if this is true or not, I just made it up for the example). Note that in this case, 4.56, 11.237, or .444 are legitimate values for the variable to assume. An important distinction between the variable in this last example and the variables in the first two examples is that the former two variables can take on only a finite number of possibilities (6 in the first case, 365 in the second), whereas by allowing fractional values (equivalently, real number values), there are an infinite number of possibilities for the variable. Note that this does not mean that a particular variable can take on any value; 12.5, 87.2, −11.0 are all possible high temperatures (Fahrenheit) for yesterday in Chicago, but −324.7 clearly is not.

A random variable is a variable that can take on one of an allowed set of values (finite or infinite in number); the actual value is determined by a happening or happenings that are not only outside our control but also are outside of any recognized, quantifiable, control; but often does seem to follow some sort of pattern.

A random variable cannot be any number, it must be chosen out of the set of possible occurrences of the situation at hand. For example, tossing a die and looking at the number that lands facing up will give us one of the variables {1,2,3,4,5,6}, but never 7, 3.2, or −4.

The good example of a simple random variable is the outcome of the flip of a coin. We can assign the number −1 to a tail and +1 to a head. The flip of the coin must yield one of the two chosen values for the random variable; but we have no way of predicting which value it will yield for a specific flip.

Is the result of the flip of a coin truly unpredictable? Theoretically, no. If you carefully analyzed the weight and shape of the coin and then tracked the exact motion of the flipper's wrist and fingers, along with the air currents present and the nature of the surface that the coin lands on, you would see that the flipping of a coin is a totally predictable event. However, since it is so difficult to track or control all these subtle factors carefully enough in normal circumstances and these factors are extremely difficult to duplicate from flip to flip, the outcome of a coin flip can reasonably be considered to be a random event. You can easily list all the possible values of the random variable assigned to the outcome of the coin flip (–1 or 1). If you believe that the coin is fair you will predict that either result is equally likely. The latter situation isn't always the case.

If you roll two dice and define the random variable to be the sum of the numbers you get from each die, then this random variable can take on any value from 2 to 12. However, all of the possible results are no longer equally likely. This assertion can be understood by looking at every possible result.

As may be seen in Table 1.1, there is only one way that the random variable can take on the value 2: Both dice have to land with a 1 face up. However, there are 3 ways that the random variable can take on the value 4: One way is for the first die to land with a 1 face up while the second die lands with a 3 face up. To avoid writing this out over and over again, I'll call this case {1,3}. By searching through the table, we see that the random variable value of 4 can be obtained by the dice combinations {1,3}, {2,2}, and {3,1}.

Table 1.2 tabulates the values of the random variable and the number of ways that each value can result from the rolling of a pair of dice.

The numbers in the right hand column of Table 1.2 add up to 36. This is just a restatement of the fact that there are 36 possible outcomes possible when rolling a pair of dice.

Define the probability of a random event occurring as the number of ways that this event can occur divided by the number of all possible events. Adding a third column to Table 1.2 to show these probabilities, we obtain Table 1.3.

If you want to know the probability that the sum of the numbers on two dice will be 5, the second column of this table tells us that there are 4 ways to get 5. Looking back at Table 1.1, you can see that this comes about from the possible combinations {1,4}, {2,3}, {3,2}, and {4,1}. The probability of rolling two dice and getting a (total) of 5 is therefore 4/36, sometimes called "4 chances out of 36." 4/36 is of course the same as 2/18, 1/9, and the decimal equivalent, .111.[2]

If you add up all of the numbers in the rightmost column of Table 1.3, you'll get exactly one. This will always be the case, because one is the sum of the probabilities of all possible events. This is the certain event and it must happen, i.e. it has a probability of one (sometimes stated as 100%).

---

2 Many fractions, such as 1/9, 1/3, and 1/6 do not have exact decimal representations that can be expressed in a finite number of digits. 1/9, for example, is .111111111…, with the ones going on forever. Saying that the decimal equivalent of 1/9 is .111 is therefore an approximation. Knowing how many digits are necessary to achieve a satisfactory approximation is context dependent, there is no easy rule.

Table 1.1 Every possible result of rolling two dice.

First die result	Second die result	Random variable value = sum of 1st and 2nd results
1	1	2
1	2	3
1	3	4
1	4	5
1	5	6
1	6	4
2	1	3
2	2	4
2	3	5
2	4	6
2	5	7
2	6	8
3	1	4
3	2	5
3	3	6
3	4	7
3	5	8
3	6	9
4	1	5
4	2	6
4	3	4
4	4	8
4	5	9
4	6	10
5	1	6
5	2	7
5	3	8
5	4	9
5	5	10
5	6	11
6	1	7
6	2	8
6	3	9
6	4	10
6	5	11
6	6	12

Table 1.2 Number of ways of obtaining each possible value when rolling two dice.

Value of random variable	Number of ways of obtaining value
2	1
3	2
4	3
5	4
6	5
7	6
8	5
9	4
10	3
11	2
12	1

Table 1.3 Table 1.2 extended to include probabilities.

Value of random variable	Number of ways of obtaining value	Probability of obtaining value
2	1	1/36 = .028
3	2	2/36 = .056
4	3	3/36 = .083
5	4	4/36 = .111
6	5	5/36 = .139
7	6	6/36 = .167
8	5	5/36 = .139
9	4	4/36 = .111
10	3	3/36 = .083
11	2	2/36 = .056
12	1	1/36 = .028

Sometimes it will be easier to calculate the probability of something we're interested in not happening than the probability of it happening. Since we know that the sum of the probabilities of our event either happening and not happening must be one, then the probability of the event happening is simply one minus the probability of the event not happening.

From Table 1.3, you can also calculate combinations of these probabilities. For example, the probability of getting a sum of *at least ten* is just the probability of getting ten + the probability of getting eleven + the probability of getting twelve, = .083 + .056 + .028 = .167. Going forward, just for convenience, we'll use the shorthand notation P(12) to mean "the probability of getting twelve," and we'll leave some things to the context, i.e. when rolling a pair of dice we'll assume that we're always interested in the sum of the two numbers facing up, and just refer to the number.[3]

Exactly what the probability of an event occurring really means is a very difficult and subtle issue. Let's leave this for later on, and just work with the intuitive "If you roll a pair of dice very many times, about 1/36 of the time the random variable will be 2, about 2/36 of the time it will be 3, etc." Since there will actually be more than one definition used, we'll note that this is the "frequentist probability" that we've introduced here.

An alternative way of discussing probabilities that is popular at horse races, among other places, is called *the odds* of something happening. This is just another way of stating things. If the probability of an event is 1/36, then we say that the odds of the event happening is 1−35 (usually written as the ratio 1 : 35). If the probability is 6/36, then the odds are 6 : 30 or 1 : 5, etc. While the probability is the number of ways that a given event can occur divided by the total number of possible events, the odds is just the ratio of the number of ways that a given event can occur to the number of ways that it can't occur. It's another way of expressing the same calculation; neither system tells you any more or any less.

In the simple coin flip game, the probability of winning equals the probability of losing, .5. The odds in this case is simply 1 : 1, often called even odds. Another term of art is the case when your probability of winning is something like 1 : 1000. It's very unlikely that you'll win; these are called long odds. While even odds is specifically defined, long odds needs some context.

You've probably noticed by now that there is no consistency about whether a probability is expressed as a fraction (such as ¼) or as a decimal (.25). Mathematically, it doesn't matter. The choice is based on context – if it seems relevant to emphasize the origins of the numerator and denominator (such as one chance out of four), a fraction will be used. When the number is either the result of a calculation or is needed for further calculations, the decimal will be used. The goal here is to enhance clarity of the point being made; insofar the mathematics is concerned, it doesn't matter.

You now have the definitions required to examine some examples. The examples to follow will start simple and then become increasingly involved.

---

3 This notation is called "functional notation." Unfortunately, it's the same notation used to show the multiplication of some variable *P* by twelve. You just have to pull the meaning out of the context. This will be discussed further in later chapters.

The goal is to illustrate aspects of the issues involved in organizing some probabilistic data and then drawing the correct conclusion(s). Examples of statistical inference will be left for later chapters.

The Lottery {Very Improbable Events and Very Large Data Sets}

Suppose you were told that there is a probability of 1 in 200 million (that's .000000005 as a decimal) of you getting hit by a car and being seriously injured or even killed if you leave your house today. Should you worry about this and cancel your plans for the day? Unless you really don't have a very firm grip on reality, the answer is clearly "no." There are (nonzero) probabilities that the next meal you eat will poison you, that the next time you take a walk it will start storming and you'll be hit by lightning, that you'll trip on your way to the bathroom and split your skull on something while falling, that an airplane will fall out of the sky and crash through your roof, etc. Just knowing that you and your acquaintances typically do make it through the day is anecdotal evidence that the sum of all of these probabilities can't be a very large number. Looking at your city's accidental death rate as a fraction of the total population gives you a pretty realistic estimate of the sum of these probabilities. If you let your plans for your life be compromised by every extremely small probability of something going wrong, then you will be totally paralyzed.[4] 1 in 200 million, when it's the probability of something bad happening to you, might as well be zero.

What about the same probability of something good happening to you? Let's say you have a lottery ticket, along with 199 999 999 other people. One of you is going to win the grand prize. Should you quit your job and order a new car based on your chance of winning?

The way to arrive at an answer to this question is to calculate a number called the expected value (of your winnings). Expected value will be defined carefully in Chapter 2. For now we'll use an intuitive "what should I expect to win?" There are four numbers we need in order to perform this calculation.

First, we need the probability of winning. In this case it's 1 in 200 million, or .000000005. Next, we need the probability of losing. Since the probability of losing plus the probability of winning must equal 1, the probability of losing must be $1 - .000000005 = .999999995$.

---

4 In 1976, when the US Skylab satellite fell from the sky, there were companies selling Skylab insurance – coverage in case you or your home got hit. If you consider the probability of this happening as approximately the size of the satellite divided by the surface area of the earth, you'll see why many fortunes have been made based on the truism that "there's a sucker born every minute."

We also need the amount of money you will make if you win. If you buy a lottery ticket for $1 and you will get $50 000 000 if you win, this is $50 000 000 − $1 = $49 999 999.

Lastly, we need the amount of money you will lose if you don't win. This is the dollar you spent to buy the lottery ticket. Let's adopt the sign convention that winnings are a positive number and losses are a negative number. The amount you'll lose is therefore −1$.

In order to calculate the expected value of your winnings, add up the product of each of the possible money transfers (winning and losing) multiplied by the probability of this event. Gathering together the numbers from above,

$$EV = (.000000005)(\$49\,999\,999) - (.999999995)(\$1)$$
$$\approx \$.25 - \$1.00 = -\$.75 \tag{1.1}$$

The expected value of your winnings is a negative number. You should expect to lose money. What the expected value is actually telling you is that if you had bought ALL of the lottery tickets, so that you had to be the winner, you would still lose money. It's no wonder that people who routinely calculate the value of investments and gambling games often refer to a lottery as a Tax On Stupidity.

The conclusion so far seems to be that events with extremely low probabilities simply don't happen. If you're waiting to win the lottery, then this is a pretty reasonable conclusion. However, the day after the lottery drawing there will be an article in the newspaper about the lottery, and a picture of a very happy person holding up a winning lottery ticket. This person just won (one dollar short of) 50 million dollars!

Can two different conclusions be drawn from the same set of data? Are we saying both that (almost) nobody wins the lottery and that somebody always wins the lottery? The answer is that there is no contradiction, we just have to be very careful how we say what we say. Let's construct an example. Suppose the state has a lottery with the probability of any one ticket winning = .000000005 and the state sells 200 million tickets which include every possible choice of numbers. It's an absolute certainty that somebody will win (we'll ignore the possibility that the winning ticket got accidentally tossed into the garbage). This does not at all contradict the statement that it's "pretty darned near" certain that you won't win.

What we are struggling with here is the headache of dealing with a very improbable event juxtaposed on a situation where there are a huge number of opportunities for this event to happen. It's perfectly reasonable to be assured that something will never happen to you while you know that it will happen to somebody. Rare diseases are an example of this phenomenon. You shouldn't spend much time worrying about a disease that randomly afflicts one person in, say, ten million, every year. But in the United States alone there will be about 35 cases of this disease reported every year. From a public health point of view, somebody should be paying attention to this disease.

A similar situation arises when looking at the probability of an electrical appliance left plugged in on your countertop starting a fire. Let's say that this probability is about 25 out of a million per person.[5] Should you meticulously unplug all your countertop kitchen appliances when you're not using them? Based on the above probability, the answer is "don't bother." However, what if you're the senior fire department safety officer for New York City, a city with about 8 million residents? Assume an average of about 4 people per residence. If nobody unplugs their appliances, then you're looking at about 50 unnecessary fires every year, possible loss of life and certain destruction of property. You are certainly going to tell people to unplug their appliances. This is a situation where the mathematics might not lead to the right answer, assuming that there even is a right answer. You'll have to draw your own conclusion here.

One last note about state lotteries. In the above example, the state took in $200 million and gave out $50 million. In principle, this is why state lotteries are held. The state makes money that it uses for education or for health care programs for the needy, etc. From this perspective, buying a lottery ticket is both a social donation and a bit of entertainment for you – that's not a bad deal for a dollar or two. On the other hand, as an investment this not only has an absurdly low chance of paying off, but since the expected value of the payoff is negative, this is not what's called a Fair Game. From an investment point of view this is a very poor place to put your money.

Coin Flipping {Fair Games, Looking Backward for Insight}

Let's set up a really simple coin flipping game between you and a friend. One of you flips a coin. If the result is heads you collect a dollar from your friend; if the result is tails you pay a dollar to your friend. Assuming that the coin is fair (not weighted, etc.), there is an equal probability of getting either a head or a tail. Since the sum of all the probabilities must equal one, then the probability of getting a head and the probability of getting a tail must be equal to .5 (one-half).

---

5 The National Fire Prevention Association (US) reported approximately 45 000 electrical fires, resulting in 420 civilian deaths, in the years 2010–2014. This is about 9000 fires per year, or about 25 fires per million people.

Letting +$1 be the result of winning (you get to be a dollar richer) and –$1 be the result of losing (you get to be a dollar poorer), then the expected value of your return is

$$E = (.5)(+1\$) + (0.5)(-1\$) = 0$$

$E = 0$ is what we're calling a fair game. Since we're defining positive $ values as winning (money coming to you) and negative $ values as losing (money leaving you), if your winnings and your losings exactly balance out, the algebraic sum is zero. Nobody involved (in this case there is just you and your friend) should expect to win, or lose, money. This last sentence seems at first to be very simple. However, examining the nuances of what is meant by expected thoroughly will take some discussion.

While Expected Value is a mathematical term that is carefully defined, the definition is not quite the same as the common, conversational, use of the term *expected* or *expectation*. Also, as a few examples will show, even in a fair game in most cases it is much less likely that you will come away from this game with the exact amount of money that you went into the game with than that you either win or lose some money.

The simplest example of this last claim is a game where you flip the coin once and then stop playing. In this case you must either win or lose a dollar – there are no other choices. For that matter, if the game is set up so that there will be any odd number of coin flips, it is impossible for you to come away neither winning nor losing. In these cases it is impossible to win the expected amount of winning – clearly a distinction between the mathematical definition of expected value and its common usage.

Now let's look at some games where there are an even number of coin flips. In these cases it certainly is possible to end up with a zero sum. The simplest example is the two-flip game. If you flip first a head and then a tail (or vice versa), you come away with a zero sum.

As a brief aside, we'll introduce some mathematical notation that will make things easier as we proceed. If we let the letter "n" refer to the number of coin flips, then our two-flip game can be referred to as an $n = 2$ game. The choice of the letter n was completely arbitrary. No new ideas or calculations have just been presented. This is simply a convenient way to talk about things. At this point there is very little to be gained by doing this, but the approach will prove to be very powerful and useful when things get more complicated.

Returning to the examination of coin flipping, let's examine the $n = 2$ game in detail. This is easy to do because there are only four possible scenarios. The results of the flips (every possible combination of heads and tails) and the algebraic values of the random variable associated with these results are shown in Table 1.4.

Table 1.4 All possible sum results in $n = 2$ simple coin flip game.

1st Flip	1st Flip variable	2nd Flip	2nd Flip variable	Sum
Head	+1	Tail	−1	0
Head	+1	Head	+1	2
Tail	−1	Tail	−1	−2
Tail	−1	Head	+1	0

As this table shows, there are two opportunities for a zero sum, one opportunity for a positive sum, and one opportunity for a negative sum. The probability that you will win is therefore .25, the probability that you'll lose is .25, and the probability that you'll break-even is .5. Note that it's equally likely that there will be a zero sum and that there will be a nonzero sum.

Now let's look at a game with a larger value of n. What about, say, $n = 10$? How many table entries must there be? For each flip of the coin, there are two possibilities. Therefore, for $n = 2$ there are four possibilities, for $n = 3$ there are $(2)(2)(2) = 8$ possibilities, for $n = 4$ there are $(2)(2)(2)(2) = 16$ possibilities, etc. Adding to our mathematical notation toolbox, the expression 2^n means 2 multiplied by itself n times. In other words, $2^3 = 8$, $2^4 = 16$, etc. We can therefore say that there are 2^n possibilities for an n-flip coin flip game.

For an $n = 10$ game,

$$2^{10} = 1024 \tag{1.2}$$

There is no reason why we can't create the list and examine every possible outcome, but it certainly does not look like doing it would be fun. Let's set our goals a little more humbly, look at $n = 3$ and $n = 4$ games. Adding to our notational soup mix, let the letter k refer to a particular flip, i.e. $k = 2$ refers to the 2nd flip, $k = 3$ refers to the 3rd flip, etc. Clearly, in an nth order game, k can take on all of the (integer) values from 1 to n. In algebraic notation, this is written as

$$1 \le k \le n \tag{1.3}$$

The symbol $\le$ means "less than or equal to," so the above expression is read as "one is equal to or less than k, and also k is equal to or less than n." This turns out to be a very convenient way to express things.

Using our new notational ability, Table 1.5 shows the $n = 3$ game.

As expected, there is absolutely no way to play an $n = 3$ game and come out even, you have to either win or lose some amount. However, since each of the outcomes (rows) in the above table is equally likely, and there are 8 of them, the

Table 1.5 All possible sum results in $n = 3$ simple coin flip game.

$k = 1$	$k = 2$	$k = 3$	Sum
−1	−1	−1	−3
−1	−1	−1	−2
−1	+1	−1	−1
−1	+1	−1	+1
+1	−1	+1	−1
+1	−1	+1	+1
+1	+1	+1	+2
+1	+1	+1	+3

probability of each outcome is exactly 1/8, so that the expected value of your return is

$$\frac{-3}{8} + \frac{-2}{8} + \frac{-1}{8} + \frac{+1}{8} + \frac{-1}{8} + \frac{+1}{8} + \frac{+3}{8} + \frac{+3}{8} = \frac{0}{8} = 0 \tag{1.4}$$

The expected value is zero, so this is indeed a fair game. Worth noting again in this case is the fact that the expected value is a value which you can never realize. You can play $n = 3$ games all night, and you will never come away with a zero (no loss or no gain) result from one of these games. What you might expect, however, is that after a night of $n = 3$ games, your net result would be close to zero.

In this simple coin flip situation, a hundred $n = 3$ games is exactly the same as an $n = 300$ game. We can therefore extend our logic: For any number of $n =$ anything games, if the total number of coin flips is odd, you can still never walk away with a zero sum. You might, however, come close.

Let's look at an $n = 4$ game and see what happens.

In an $n = 4$ game, there are $2^4 = 16$ possible outcomes, all listed in Table 1.6. Looking through the column of sums, we see that there are 6 possible ways to get 0. In other words, the probability of neither winning nor losing is 6/16 = .375. This is lower than the probability of neither winning nor losing was for an $n = 2$ game. Is this a pattern?

Table 1.7 shows that the probability of getting a zero sum out of even order games for $2 \le n \le 20$.

There is indeed a trend – as n gets larger, the probability of getting a zero sum gets lower. The more times you flip the coin, the less likely it is that you will get exactly the same number of heads and tails.

Table 1.6 All possible sum results in $n = 4$ simple coin flip game.

$k = 1$	$k = 2$	$k = 3$	$k = 4$	Sum
−1	−1	−1	−1	−4
−1	−1	−1	+1	−2
−1	−1	+1	−1	−2
−1	−1	+1	+1	0
−1	+1	−1	−1	−2
−1	+1	−1	+1	0
−1	+1	+1	−1	0
−1	+1	+1	+1	2
+1	−1	−1	−1	−2
+1	−1	−1	+1	0
+1	−1	+1	−1	0
+1	−1	+1	+1	2
+1	+1	−1	−1	0
+1	+1	−1	+1	2
+1	+1	+1	−1	2
+1	+1	+1	+1	4

Table 1.7 Probabilities of zero sum in simple coin flip game.

n	P(zero sum)
2	.500
4	.375
6	.313
8	.273
10	.246
12	.226
14	.209
16	.196
18	.185
20	.176

This leads to some very thought-provoking discussions of just what a probability means, what you can expect, and how sure you can be of what you expect. This in turn leads to a very basic question of just what it means to be sure about something, or in other words, just how confident you are about something happening. In Chapter 3 we'll define a confidence factor by which we can gauge how sure we are about something.

Returning to the simple coin flip, what if we want to know the probability of getting five heads in a row? This is identically the probability of flipping five coins and getting five heads, because the results of coin flips, be they with multiple coins or with the same coin over and over again are independent of each other. You could get the answer by writing a list such as the tables I've been presenting for the $n = 5$ case or you could just note that for independent events, all you have to do is to multiply the individual probabilities together. In other words, the probability of getting five heads in a row is

$$\left(\frac{1}{2}\right)^5 = \frac{1}{32} \approx .033.$$

.033 is a low probability, but not so low as to astound you if it happens. What if it indeed just did happen? You flipped a coin and got five heads in a row. What's the probability of getting a head if you now flip the coin again? Assuming, of course, that the coin isn't weighted or corrupted in any other manner, i.e. that the flips are indeed fair, the probability of a head on this flip (and for any subsequent flip) is still just ½. Putting it simply, a coin has no memory.

Reiterating the point above, flipping one coin six times is statistically identical to flipping six different coins once each and then examining the results. It doesn't matter whether you flip the six coins one-at-a-time or if you toss them all up into the air and let them fall onto the table. The six coins are independent of each other, they do not know or remember anything about either their own past performance or the performance of any other coin. When you look at it this way, it's pretty clear that the flip of the sixth coin has nothing to do with the flips of the first five coins. For that matter, if you tossed all six coins into the air at once, you couldn't even say which coin is the sixth coin.

The term independent was used several times above without stopping to discuss it. Independent events are events that have nothing to do with each other and don't affect each other. If I have a hat full of numbered pieces of paper, say one to a hundred, the probability of pulling (the piece of paper numbered) 50 is one out of a hundred. If I put the piece of paper back and then pull another number out, these (pullings) are independent events. On the other hand, if I don't replace the number after pulling it out and then pull another number, these events are not independent because the first event left the situation changed for the second event.

The above arguments are a simple case of another interesting discussion. When does knowledge of past results tell you something about future results?

In the above example, it doesn't tell you anything at all. Later in this chapter we will have an example where this isn't the case.[6]

Returning to the coin flipping game, remember that the expected value of return of an *n* equals anything coin flip is always zero, as has been illustrated in several examples. If you were to flip a coin ten times, the most likely single result would be an equal number of heads and tails, even though it's not a very likely event (remember, in this case we're counting the number of ways of getting five heads and five tails out of 2^{10} = 1024 possible configurations). The distinction between most likely and yet not very likely (or the equivalent, very unlikely), eludes many people, so let's consider another example.

Suppose we have a giant roulette wheel with 1000 slots for the ball to land in. Number the slots 1–999 consecutively, and then number the thousandth slot 500. This means there is exactly one slot for each number between 1 and 999, except for the number 500, for which there are two slots. Spin this roulette wheel and watch for the ball to settle in a slot; you see that there are two opportunities for the ball to settle in a slot numbered 500, but only one opportunity for the ball to settle at any other number. In other words, the probability of the ball landing at 500 is twice the probability of the ball landing at any other number. 500 is clearly the most likely result. The probability of the ball landing in a 500 slot is 2 out of 1000, or .002. The probability of the ball not landing in a 500 slot, that is, the probability of landing anywhere but in a 500 slot is 998 out of 1000, or .998. It is, therefore, very unlikely that the ball will land in a 500 slot. Now let's combine both of the above observations into the same sentence: the most likely slot the ball will land in is numbered 500, but it is very unlikely that the ball will land in a slot numbered 500 (as compared to some other slot).

Returning to the coin flipping example, no matter how many (even number of) times you flip a coin, the most likely result is that you'll get an equal number of heads and tails. The more times you flip the coin, however, the less likely this result will be. This same idea will be presented in the chapter on random walks.

A variation on the above example is picking a number for a lottery. If you were to pick, say, 12345, or 22222, you would be criticized by the "experts": "You never see a winning number with a regular pattern. It's always something like 13557 or 25738 or...." This last statement is correct. It is correct because of all the nearly one million five digit numbers that can be picked, very few of them have simple, recognizable digit patterns. It is therefore most likely that the winning number will not have a recognizable pattern. However, the five lottery balls have no memory or awareness of each other. They would not know if they presented a recognizable pattern; actually we would be hard pressed to define what is meant by a "recognizable pattern." Any five digit

---

6 People who don't understand this point will tell you that if you flipped a coin 100 times and got 100 heads, you should bet on a tail for the 101st flip because "it's due." More likely, this is not a fair coin, it "wants" to flip heads. After seeing 100 heads in a row, it's a very good bet that the 101th flip will also be a head. This will be discussed later in the Bayes Theorem chapters.

number is equally likely to be picked. The difference between the recognizable patterns and other patterns is only in the eyes of the beholder.

A corollary to this is that there's no reason not to pick the very number that won last week. It's highly unlikely that this number will win again just because there are so many numbers to choose from, but it is just as likely that this number will win as it is that any other number will be win.

Moving on, what if you have just flipped a coin five times, got five heads, and now want to flip the coin ten times more? The expectation value looking forward is still zero. But, having just won the game five times, you have five dollars more in your pocket than you started with. Therefore, the most likely scenario is that you will end up with five dollars in your pocket! This property will be gone into in more detail in the chapter on gambling games. Generalizing this conclusion, we can say that if you are going to spend the evening flipping coins (using these rules), your most likely status at the finish is just your status at the time you think about it. If you start off lucky (i.e. have net winnings early on) then you'll probably end up winning a bit, and vice versa. There really is such a thing as being on a winning streak. However, this observation can only be correctly made after the fact. If you were lucky and got more heads than tails (or vice versa, if that's the side you're on) then you were indeed On A Winning Streak. The perception that being on a winning streak will influence the coin's future results is of course total nonsense. You might win a few times in a row, you might even win most of the time over the course of the evening, but each flip is still independent of all the others.

There is an argument for saying that if you have been on a winning streak it's more likely that you'll end the evening ahead (i.e. with net winnings rather than losses) than if you haven't been on a winning streak. This argument is that if you have been on a winning streak since you started the evening you have a lot more money in your pocket than you would have if you had been on a losing streak. You are therefore in a better position to withstand a few (more) losses without being wiped out and having to quit playing, and therefore your odds of winning for the evening have been increased. This argument has nothing to do with the probabilities of an individual win (coin flip, roulette wheel, poker hand, whatever). If you are just playing for a score on a piece of paper and cannot be wiped out, this argument is pointless.

The Coin Flip Strategy That Can't Lose

Assume that you want to earn $1 a day. Let's build a strategy for playing a clever coin flipping game:

1) Bet $1 on the results of a coin flip.
2) If you win the 1st coin flip, you've earned your $1 for the day. Go home.
3) If you lose the 1st coin flip, you've lost $1. Bet $2 and try again.

4) If you win the 2nd coin flip, then you've recovered your lost $1 and won $1. Go home.
5) If you lose the 2nd coin flip, then you've now lost $3. Bet $4 and try again.
6) If you win the 3rd coin flip, then you've recovered your lost $7 and won $1. Go home.
7) If you lose the 3rd coin flip, then you've now lost $7. Bet $8 and try again.
8) Continue until you win.

This scheme seems unbeatable. If you keep flipping a coin, sooner or later you have to get a head, and you've won for the day. What could possibly go wrong?

The scheme could be analyzed in detail from a number of different perspectives. The Achilles' heel is that you need start with a very full wallet. For example, to cover 3 losses and still have the money to place your 4th bet, you need to start with $1 + $2 + $4 + $8 = $15. In general, to be able to place n bets you need to start out with $2^n - 1$ dollars.

If you start the evening with one million dollars, it's pretty certain that you will be able to go home with $1 million + $1. You have enough money to cover seeing a lot of tails until you get your head. On the other hand, if you show up with only $1, then the probability of you going home with your starting $1 plus your winnings of $1 is only 50%. Putting this last case slightly differently, the probability of you doubling your money before getting wiped out is 50%. Without showing the details now, it turns out that no matter how much money you start with, the probability of you doubling your money before you get wiped out is at best 50%, less if you have to flip the coin many times. If you're "earning" only a dollar a day, then you need to come back the number of days equal to the number of dollars you're starting with. However, you can save yourself a lot of time by just betting all of your money on the first coin flip – a very simple game with a 50% probability of doubling your money and a 50% probability of being wiped out. Here again, we have been reconciling an unlikely event (not getting a head after many flips of a coin) with a large number of opportunities for the event to happen (many flips of a coin). It's not possible to fool Mother Nature. If you start out with a million dollars and plan to visit 10 000 gambling houses each night, hoping to win only $1 at each house, then the probabilities start catching up with you and you no longer have a sure thing going.

The Prize Behind the Door {Looking Backward for Insight, Again}

This example is subtle, and the answer is often incorrectly guessed by people who should know better. There are still ongoing debates about this puzzle on various Probability Puzzle websites because the correct answer seems to be so counter-intuitive to some people that they just won't accept the analyses. It is known by many names, most commonly the Monty Hall problem, after the host of a TV game show.

You are a participant in a TV game show. There are three doors (let's call them doors A, B, and C). Behind one of these doors is a substantial prize, behind the other two doors is nothing. You have to take a guess. So far this is very straightforward, your probability of guessing correctly and winning the prize is exactly 1/3.

You take (and announce) your guess. Before the three doors are opened to reveal the location of the prize, the game show host goes to one of the two doors that you didn't choose, opens it, and shows you that the prize is not behind this door. The prize, therefore, must either be behind the unopened door that you chose or the unopened door that you did not choose. You are now given the option of staying with your original choice or switching to the unopened door that you did not choose. What should you do?

Almost everybody's first response to this puzzle is to shrug – after all, there are now two unopened doors and the prize is behind one of them. Shouldn't there simply be a .5 (50%) probability of the prize being behind either of these doors, and therefore it doesn't matter whether you stay with your original choice or switch?

Let's look at all the possible scenarios. Assume that that your first guess is door B. (It doesn't matter which door you guess first, the answer always comes out the same.)

1) If the prize is behind door A, then the host must tell you that the prize is not behind door C.
2) If the prize is behind door B, then the host can tell you either that the prize is not behind door A or that the prize is not behind door C.
3) If the prize is behind door C, then the host must tell you that the prize is not behind door A.

Since each of the above three situations is equally likely, they each have a probability of 1/3.

Situation 1: you stay with your first choice (door B), you lose. You have the option of switching to door A. If you switch to door A, you win.
Situation 2: you stay with your first choice (door B), you win. If the host tells you that the prize is not behind door A and you switch to door C, you lose. Also, if the host tells you that the prize is not behind door C and you switch to door A, you lose.
Situation 3: you stay with your first choice (door B), you lose. You have the option of switching to door C. If you switch to door C, you win.

At this point, Table 1.8 is in order. Remember, your first choice was door B.

Table 1.8 Monty hall game door choice alternatives.

Prize location	Remaining doors	Definite losers	Stay with B	Switch
A	A, C	C	LOSE	WIN
B	A, C	A and C	WIN	LOSE
C	A, C	A	LOSE	WIN

If you stay with your first choice, you only win in one of three equally likely situations, therefore your probability of winning is exactly 1/3. This shouldn't really surprise you. The probability of correctly guessing one door out of three is 1/3, and there's not much more that you can say about it.

On the other hand, if your only options are to stay with your first choice or to switch to the other unopened door, then your probability of winning if you switch must be 1 − 1/3 = 2/3. There's no getting around this: either you win or you lose and the probability of winning plus the probability of losing must add up to the certain event, that is, to a probability of 1.

What just happened? What has happened that makes this different than having just flipped a coin five times, having gotten five heads, and wondering about the sixth flip?

In the coin flipping example, neither the probabilities of the different possibilities or your knowledge of these probabilities changed after five coin flips. In other words, you neither changed the situation nor learned more about the situation. (Obviously, if someone took away the original coin and replaced it with a two-headed coin, then expectation values for future flips would change.) In the game-show example, only your knowledge of the probabilities changed; you learned that the probability of the prize being behind one specific door was zero. However, this is enough to make it possible that the expected results of different actions on your part will also change.

In a later chapter we'll look at another very unintuitive situation. In a combination of games, known as "Parrondo's paradox" – jumping randomly between two losing games creates a winning game because one of the losing games involves looking back at how much you've already won or lost.

The Checker Board {Dealing With Only Part of the Data Set}

Imagine an enormous checker board: the board is 2000 squares by 2000 squares. There are 2000 × 2000 = 4 000 000 (four million) squares on the board. Assume that each square has an indentation that can capture a marble.

Treat the board as an imaginary map, and divide it up into regions, each region containing 1000 indentations. The regions themselves need not be square or even rectangular, so long as each region contains exactly 1000 indentations. There are 4 000 000/1000 = 4000 of these regions on the board. There is nothing specific in the choice of any of these numbers. For the purposes of the example, we need a large total area (in this case four million squares) divided into many small regions (in this case four thousand regions). Also, the regions do not all have to be the same size, making them the same size just makes the example easier to present.

Now, lay this checkerboard flat on the ground and then climb up to the roof of a nearby building. The building must be tall enough so that the checkerboard looks like a small dot when looking down. This is important because it assures that if you were to toss a marble off the roof it would land somewhere on or near the checkerboard, but you can't control where. Start tossing marbles off the roof until 40 000 marbles have landed on the checkerboard and are trapped in 40 000 indentations. This is, admittedly, a very impractical experiment. Don't worry about that; we're not really planning to perform this experiment. It's only a way to describe a scattering of 40 000 objects into 4 000 000 possible locations. The choice of 40 000 objects isn't even a critical choice, it was just important to choose a number that is a small fraction of 4 000 000 but is still a fairly large number. In any case, the fraction of the number of indentations that are filled is exactly

$$\frac{40\,000}{4\,000\,000} = \frac{1}{100} = 1\% \tag{1.5}$$

Now let's take a close look at a few of the 4000 regions, each of which has 1000 indentations. Since 1% of the indentations are filled with marbles, we would expect to see 1% of 1000, or .01 × 1000 = 10 marbles in each region. On the average, over the 4000 regions, this is exactly what we must see – otherwise the total number of marbles would not be 40 000. However, when we start looking closely, we see something very interesting[7]: Only about 500 of the regions have 10 marbles.[8] About 200 of the regions have 14 marbles, and about 7 of the regions have 20 marbles. Also, about 9 of the regions have only 2 marbles. These results are tabulated in Table 1.9.

---

7 At this point we are not attempting to explain how the observations of "what we actually see" come about. This will be the subject of the chapter on binomial distributions.

8 We say "about" because it is very unlikely that these numbers will repeat exactly if we were to clear the board and repeat the experiment. How the results should be expected to vary over repeated experiments will also be the subject of a later chapter.

Table 1.9 shows us that the most likely situation is in this case 10 marbles per region; this will happen more often than any other situation. But, the most likely situation is not the only thing that can happen (just as in the coin flip game). The results are distributed over many different situations, with less likely situations happening less often. Therefore, in order to predict what we see from this experiment, we not only need to know the most likely result, but we also need to know something about how a group of results will be distributed among all possible results. These probability distributions will be a subject of the next chapter.

Since the most likely result (10 marbles per region) occurs only about 500 times out of 4000 opportunities, some other results must be occurring

Table 1.9 Estimated number of marbles in regions.

Nr of regions	Nr of marbles
2	1
9	2
30	3
74	4
150	5
251	6
360	7
451	8
502	9
503	10
457	11
381	12
292	13
208	14
138	15
86	16
50	17
28	18
14	19
7	20
3	21
2	22
1	23

about 3500 times out of these 4000 opportunities. Again, we have to be very careful of what we mean by the term "most likely result." We mean the result that will probably occur more times than any other result when we look at the whole checkerboard. The probability that the most likely result will not occur in any given region is about

$$\frac{3500}{4000} = \frac{7}{8} = .875 = 87.5\% \tag{1.6}$$

In gambling terms, there are 7 to 1 odds against the most likely result occurring in a given region.

Now, suppose someone is interested in the regions that have at least 20 marbles. From the table, we see that there are 13 of these regions. It wouldn't be surprising if a few of them are near an edge of the board. Let's imagine that this person locates these regions and takes a picture of each of them. If these pictures were shown to you and you are not able to look over the rest of the board yourself, you could believe the argument that since there are some regions near the edge of the board that have at least twice the average number of marbles, then there must be something about being near the edge of the board that attracts marbles. There are, of course, many regions that have less than half the average number of marbles, and some of these are probably near the edge of the board too, but this information is rarely mentioned (it just doesn't make for good headlines). Instead, we see "Cancer Cluster Near Power Lines" and similar statements in the newspapers. It's hard to generalize as to whether the reporters who wrote the story intentionally ignored some data, unintentionally overlooked some data, didn't understand what they were doing by not looking at all the data, or were just looking to write a good story at the cost of total, complete, truthfulness.

In all fairness, there are unfortunate histories of true disease clusters near waste dumps, etc. However, the point that must be made over and over again is that you cannot correctly spot a pattern by selectively picking (sometime referred to a cherry picking) subsets of a large data set. In the case of power lines near towns, when you look at the entire checkerboard (i.e. the entire country) you find high disease clusters offset by low disease clusters and when you add everything up, you get the average. If there were high disease clusters which were truly nonrandom, then these clusters would not be offset by low disease clusters, and you would get a higher than average disease rate when you add everything up. Also, you must be prepared to predict the variability in what you see, e.g. a slightly higher than average total might just be a random fluctuation, sort of like flipping a coin and getting 5 heads in a row.[9]

---

9 The World Health Organization maintains a database on their web site in the section on ElectroMagnetic Fields where you can get a balanced perspective of the power line and related studies.

This same issue shows up over and over again in our daily lives. We are bombarded with everything from health food store claims to astrological predictions. We are never shown the results of a large study (i.e. the entire checker board). The difficult part here is that some claim might be absolutely correct. However, the point is that we are not being shown the information necessary to see the entire picture, so we have no way of correctly concluding whether or not a claim is correct or just anecdotal, without further investigation. And, of course in the area of better health or longer life or ..., we are genetically programmed to bet on the pattern we are shown, just in case there are man-eating tigers lurking in the brush.

Comments

It is not uncommon to be confronted with a situation where it is either impossible or very highly impractical to study the entire data set. For example, if we are light bulb manufacturers and we want to learn how long our light bulbs last, we could ask every customer to track the lifetime of every bulb and report back to us, but we know that this is not going to happen. We could run all of our light bulbs ourselves until they burn out, but this is not a very good business plan. Instead, we take a representative sample of the bulbs we manufacture, run them ourselves until they burn out, and then study our data. In a later chapter, we will discuss what it takes for a sample to be representative of the entire population. We need to know what fraction of the manufactured bulbs we must choose as samples, what rules we need to follow in selecting these samples, and just how well the results of studying the sample population predict the results for the entire population.

Hopefully, these examples and explanations have created enough interest so that you want to continue reading. Chapters 2 and 3 will present most of the basic mathematics and definitions needed to continue. Then, starting in Chapter 4, we will show examples of how random variables and probability affect many of the things we do and also how they are at the basis of many of the characteristics of the world we live in.

Before moving on, we must say that the definition of probability that we've been working toward in this chapter is not the only such definition. This definition is sometimes called the frequentist definition because of the implicit assumption that it's what we would get, in some kind of average, if we repeated our experiment over and over and over, or equivalently, created many parallel universes and did our experiment in each of them. The term frequentist arises from the idea that we are calling on frequent repeats of our experiment to bring meaning to our concept of probability.

There is an entire other way of thinking about probability, sometimes called the diachronic approach. Diachronic means evolving over time, in our usage evolving as we bring in new information. In this approach, also known as the Bayesian approach, the probability of an event occurring does not refer to the event itself, but to our belief that this event will occur. We begin with some background information and estimate of this probability, and then improve our estimate as we acquire new information.

Both of these approaches will be discussed in upcoming chapters. The frequentist approach is the more common, so we will begin with it.

Problems

1.1 We have a small deck of cards. It has five cards: The ace (1) of spades and ace of hearts, 2 of spades and 2 of hearts, and a joker. The solutions will show probabilities calculated by the most basic of all techniques – list all possible events, select the event(s) which satisfy the criterion(a) of the problem, and divide the latter by the former.

In all cases the deck is shuffled before the start of the problem and cards are drawn from the top of the deck.

A What is the probability of drawing a 2?

B What is the probability of drawing a 2 or the joker?

C What is the probability of drawing a one and then a two?

D What is the probability of drawing a one and a two without caring which card comes first? In card-playing jargon, this problem is expressed as "*What is the probability of being dealt a one and a two from this deck?*" The distinction between this and part c is that in this problem the *order* of drawing the 1 and the 2 does not matter.

E Assign the joker a numerical value of 0. What is the probability of drawing two cards whose numerical values add up to 3?

F Again, assign the joker a numerical value of 0. What is the probability of drawing two cards whose numerical values add up to 1?

G What is the probability of drawing the joker, then after putting the joker back in the deck and shuffling the deck, drawing the joker again?

1.2 A When rolling a pair of dice, it is conventional to consider the result to be the sum of the numbers facing up, as was discussed above. Let's change things by considering the difference of the two numbers. Take this one step further, consider the absolute value of the difference. That is, we'll always take the larger number minus the smaller number (if both numbers are the same, the difference is zero). What is the probability of rolling the dice and getting zero?

B Continue the above problem by listing all the possible results and their probabilities.

1.3 Two coins are tossed into two boxes at random. What is the probability that one of the boxes is empty?

1.4 A drawer contains two pairs of socks – one white and one brown. Two socks are drawn at random. What is the probability that they are a matched pair?

1.5 Many gambling games traditionally speak of *odds* rather than *probabilities*. The odds of something occurring (called the odds *for* the event or the odds *in your favor*) is the probability of the event happening divided by the probability of the event not happening:

$$Odds \equiv \frac{P}{1-P}$$

Odds are typically expressed as the ratio of two integers, reduced to their lowest common denominator. For example, the odds of flipping a fair coin and getting heads is

$$Odds = \frac{.5}{1-.5} = \frac{.5}{.5} = 1:1$$

$1:1$ is sometime called *even* or *fair* odds.

A If you are betting on something where you have a .4 probability of winning, what are the odds in your favor (for you) and what are the odds against you?

B Suppose the quoted odds on a horse at a race are $5:2$. What is your probability of winning if you bet on this horse?

1.6 A bowl contains 10 red candies, 10 white candies, and 15 green candies.

A A candy is randomly chosen. What is the probability that it is red?

B A candy is randomly chosen. What is the probability that it is not red?

C A candy is randomly chosen. What is the probability that it is not white?

D Two candies are successively randomly chosen. What is the probability of choosing a red candy and then a green candy if the first candy is replaced before choosing the second candy?

E Repeat the above if the first candy is not replaced before choosing the second candy?

F Three candies are successively randomly chosen. What is the probability of choosing a red and then a green and then a white candy if the first candy is replaced before choosing the second candy but the second candy is not replaced before choosing the third candy?

1.7   You bet $4 and then roll a standard die. Your winning is the number you roll, in dollars.

    A What is the expected value of your winnings?

    B What would you have had to bet for this to be a fair game?

1.8   This is a somewhat complicated gambling game. You flip a (fair) coin. If you flip a head, you roll a die and your winnings (in dollars) is what you roll. If you flip a tail, you roll the same die and your losses are what you roll. Is this a fair game?

1.9   This is a problem that requires some knowledge of sums of infinite series to get the exact answer, but simple arithmetic can give us a very good approximation to the exact answer – a bit tedious by hand, but very easy on a spreadsheet.

    You and a friend take turns rolling a pair of dice. The first person to roll an 8 wins. Assuming you go first, what is your probability of winning?

1.10   This is a continuation of Problem 1.9. Consider the first game your first flip. Consider the second game, if there is one, your friend's first flip. Consider the third game, if there is one, your second flip. What is the probability that there will be a fourth game?

1.11   A business has prepared 3 invoices and 3 addressed letters for 3 different customers. The invoices are put into the envelopes randomly.

    A What is the probability that all 3 invoices are in their correct envelopes?

    B What is the probability that none of the invoices are in their correct envelopes?

    C What is the probability that exactly 1 of the invoices are in its correct envelope?

    D What is the probability that at least 1 invoice is in its correct envelope?

2

Probability Distribution Functions and Some Math Basics

The Probability Distribution Function

(Aside) This section includes a primer/review of the mathematical notations introduced in this chapter. The mathematics is mostly just combinations of adding, subtracting, multiplying, and dividing lists of numbers. Since – in many cases – we will be dealing with very many numbers at a time (hundreds or even thousands of them), it is impractical to write out all of the numbers involved in a given calculation. What we do instead is introduce a summarizing notation that includes things such as *subscripted variables, summation signs*.

The probability distribution function (PDF) is a powerful tool for studying and understanding probabilities. However, before discussing it is important to introduce the idea of a mathematical function. A mathematical function, or more simply, a function, is the mathematical equivalent of a food processor: you pour in one or more numbers, push the "grind" button for a few seconds, and pour out the resulting concoction.[1] Three important facts were just presented:

1) One or more numbers, usually called *variables* or *independent variables* go into the function, depending upon the particular function "recipe" we're dealing with.
2) The function somehow processes these numbers to produce a result (a number) which is usually called the *value of the function* for the particular variable(s) that went into the function.
3) The function produces exactly one value (result) for any given input variable(s).

Fact 1 is true in the general case, but the cases we will be dealing with have only one number going into the function – so from now on we'll limit our discussion to this case. As an example, consider the function "Double the number

---

1 Needless to say, this is a shameless simplification of a very important topic in mathematics.

Probably Not: Future Prediction Using Probability and Statistical Inference,
Second Edition. Lawrence N. Dworsky.

coming in and then add 3 to it." If the number coming in is 5, the result is 13; if the number coming in is 2, the result is 7, etc.

In a simple mathematical notation, the input number is represented by a letter, often x, by convention almost always letters chosen from near the end of the alphabet. x can take on many possible values; hence the name "variable," and the above function *of the variable x* is written as

$$function = 2x + 3 \tag{2.1}$$

Sometimes a function must be written as one or more rules. For example, the *absolute value* function of [the variable] x is defined by two rules:

1) If x is equal to or greater than 0, the function equals x
2) If x is less than 0 (negative), the function equals $-x$.

Examples of the absolute value function are:

- $x = 2$, function = 2
- $x = 4$, function = 4
- $x = -3$, function = 3
- $x = -6.4$, function = 6.4
- $x = 0$, function = 0

The absolute value function is a common enough function for a special notation to have been developed for it: Absolute value of $x = |x|$.

In both of the above examples, [the variable] x could have been any number that you can think of. However, sometimes the allowed values of x are restricted. For example, suppose our function is "The probability of getting a particular number which is the sum of the two faces of a pair of dice when we roll these dice." The only allowed numbers for the input variable are the possibilities that you can get when you roll a pair of dice, the integers from 2 to 12. This function was described in Chapter 1. If the input variable is 2, the function = 1/36. If the input variable is 3, the function is 2/36, etc.

Functions that describe probabilities are known as PDFs. All PDFs share several attributes:

1) The function can never be less than 0. We have no concept of what a negative probability might mean.
2) If we are dealing with a (finite length) list of numbers, the function can never be greater than 1. Remember that for a list of numbers a probability of 1 describes a certain event. A probability greater than 1 doesn't mean anything.
3) When the allowed inputs to a function can be written as a finite length list of numbers, the sum of the values of the function for all of the allowed inputs must equal 1. This situation was also described in Chapter 1. Every allowed number going into the function has associated with it a result, or value of the function. Since the list of allowed numbers covers, as its name

implies, every possible input number to the function, all of the values of the function added together (the sum of all the possible values of the function) must be a certain event (i.e. one of these events must occur): this sum must equal one.

The numbers going into a PDF are *variables* which are numbers chosen from an allowed set of numbers (the situation usually makes it very clear as to just what the allowed set of numbers is). When we invoke a particular situation, one of the allowed variables is chosen, seemingly at random, according to the likelihoods assigned by the probabilities attached to all of these [allowed] random variables.

A very useful way of looking at PDFs is to really *look* at the PDF (no pun intended) by drawing a picture, or graph (plot), of the function. Figure 2.1 is a plot of the PDF of the probabilities of the dice roll that were tabulated in Table 1.3.

The allowed values of the horizontal axis (typically "the *x* axis") are the integers from 2 to 12. If – for some reason – not all of the allowed values are shown, then the graph is only part of the PDF and should be carefully labeled as such.

The vertical (typically the "*y*") axis represents the values of the function, in this case the probabilities shown in Table 1.3; we know that the only possible numbers are between 0 and 1.

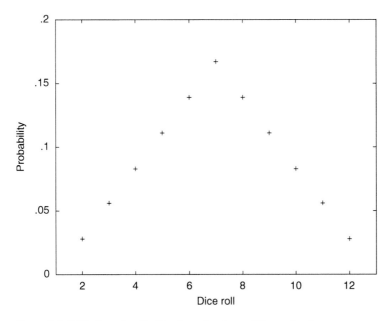

Figure 2.1 PDF of results of rolling two dice (sum of face values).

Sometimes a PDF has hundreds or possibly thousands of allowed values of x. When this happens, we often replace the hundreds of dots with a continuous line. This is just a convenience to make the graph easier to look at. This is a very important point. When the PDF is defined as a *discrete* PDF, there are allowed values of x, each represented by a point on the graph. Drawing a line between these points (and then forgetting to draw the points) does NOT mean that we allow values of x in between the x point locations. For example, a single die roll can yield values of x which are the integers from 1 to 6. 3.5 or 4.36 or any other noninteger value simply does not make sense. On the other hand, there is a category of PDFs known as *continuous* PDFs which does allow for any value of x – possibly between two limits or possibly even *any* number x. An example of this is the probability of a certain outside temperature at noon of a given day (any value of x allowed, but x values between about –60 and +125 F should cover most places where you'd consider living).

Another example is the probability of the weight or height of the next person coming through the door. The same concepts of probability and PDF hold for continuous as for discrete PDFs, but we must reconsider what we mean when we say that the "sum of the probabilities must be one." If we are to allow any value of x between, say 1 and 2, then we must allow 1, 1.1, 1.11, 1.111, 1.1111, etc. How can we add up all of these probabilities? The answer lies in the integral calculus – we learn to "add up" an infinite number of cases, each of which applies to an infinitesimally small region on the x axis. The conclusion is that we replace the requirement that the sum of the finite number of probabilities equals one with the requirement that the area under the probability curve equals one. This might be confusing now, but examples to follow should clear up the ideas.

Averages and Weighted Averages

We need to consider ways to summarize the properties of PDFs, both discrete and continuous. One way, perhaps the most important way, is with averages. At the same time, we can introduce some very useful concise notation.

Consider the list of ten numbers {1,2,3,3,3,4,5,5,5,5}. The average of these numbers, also called the *arithmetic mean* or often simply the *mean*, is just the sum of these numbers divided by how many numbers were summed:

$$\frac{1+2+3+3+3+4+5+5+5+5}{10} = \frac{36}{10} = 3.6 \tag{2.2}$$

This is a good place to introduce some notation. No new idea is being introduced here, we're just writing things in a more concise "language" that suits our needs better than words do.

In the above example, there are ten numbers in the list. The first number is 1, the second number is 2, the third number is 3, the fourth number is also 3, etc.

We'll generalize the idea of referring to a specific number in the list by adding a subscript i: x_i now refers to the "ith" number in the list. Remember that the values of i have nothing to do with what the numbers in the list are; they just tell you where in the list to look for the number:

- $x_1 = 1$
- $x_2 = 2$
- $x_6 = 4$, etc.

Part of calculating the average of this list of numbers is adding up the numbers in the list. We say that we want to add up (get the sum of) the numbers in the list by using the upper case Greek letter Sigma: Σ

The notation for the sum of the above list of numbers is

$$sum = \sum_{i=1}^{10} x_i \tag{2.3}$$

Taking Eq. (2.3) apart piece by piece, the upper case Sigma simply means that we're going to be adding something up. What are we adding up? Whatever appears to the right of the Sigma, in this case some x_i's. Which of them are we adding up? Under the Sigma is $i = 1$. That means start with x_1. On top of the Sigma is 10. This means $i = 10$ (it's not necessary to repeat the i). This means stop after 10. Putting all of this together,

$$sum = \sum_{i=1}^{10} x_i = x_1 + x_2 + \cdots + x_{10} = 1+2+3+3+3+4+5+5+5+5 = 36 \tag{2.4}$$

If we wanted to add up the second through the fourth numbers of the list, we would write

$$sum = \sum_{i=2}^{4} x_i = x_2 + x_3 + x_4 = 2+3+3 = 8 \tag{2.5}$$

The expression to the right of the sigma can itself be a function. For example,

$$sum = \sum_{i=3}^{5} \sqrt{x_i^2 - \frac{x}{2}} = \sqrt{3^2 - \frac{3}{2}} + \sqrt{4^2 - \frac{4}{2}} + \sqrt{5^2 - \frac{5}{2}} = 11.22 \tag{2.6}$$

Two comments:

1) Occasionally you'll see a Sigma with nothing above or below it. This is an informal notation which means "Look to the right, what we're adding up should be obvious." From the context of whatever you see, you should be able to identify everything that can be added up – go ahead and add them all up.
2) The capital pi symbol, Π, is used in the same way as the capital sigma. This symbol means "multiply all the terms together" instead of "add all the terms together."

Using our notational tool, the formula for the average of the list of numbers is

$$average = \frac{1}{10}\sum_{i=1}^{10} x_i \tag{2.7}$$

The general expression for the average of a list of n numbers is

$$average = \frac{1}{n}\sum_{i=1}^{n} x_i \tag{2.8}$$

This formula by itself is useless without a list of n numbers (or instructions how to create this list).

As an example, consider the list

$$x_1 = 9$$
$$x_2 = 16 \tag{2.9}$$
$$x_3 = 25$$

The average of these numbers is

$$average = \frac{1}{3}\sum_{i=1}^{3} x_i = \frac{9+16+25}{3} = 16.667 \tag{2.10}$$

Let's repeat the list of 10 numbers (Eq. (2.2)) so that we have it in front of us: {1,2,3,3,3,4,5,5,5,5}. The number 3 appears three times and the number 5 appears four times. Another way of specifying this list is by writing a shorter list of numbers showing each number that appears only once: {1,2,3,4,5}, and then creating a second list called a *Weighting List* which we'll represent with the letter w_i. The weighting list tells us how many times each of the numbers in our main list appears. In this case, the weighting list is the list {1,1,3,1,4}. Putting these together, these lists tell us that 1 appears once, 2 appears once, 3 appears three times, 4 appears once, and 5 appears four times.

We can simplify our original list by not showing the repeats. We'll call this the simplified list x_i, {1,2,3,4,5,}.

In our shorthand notation, the sum of the 10 numbers is calculated as

$$sum = \sum_{i=1}^{5} w_i x_i = (1)(1)+(1)(2)+(3)(3)+\cdots+(4)(5) = 36 \tag{2.11}$$

Note that both lists must have the same number of items (in this case 5).

The average, referred to as a *weighted average*, is written as

$$average = \frac{\sum_{i=1}^{5} w_i x_i}{\sum_{i=1}^{5} w_i} \tag{2.12}$$

In Eq. (2.12), the numerator is identically the sum as shown in Eq. (2.11). Looking back at the definition of the weighting list, the ith item in the weighting list tells us the number of times we will see w_i in the main list. If we add up the numbers of the weighting list, we must get the total number of items in the main list. Going back to the definition of average, Eq. (2.8), this is identically n (the number of items in the list). Weighted averages will appear often in pages to follow.

Expected Values (Again)

Suppose the weighting function is the PDF that is associated with the variables x_i. Before showing the uses of this, let's go through the formalities of the calculation. In terms of notation, just to remind us that in this situation the weighting list is a PDF, we'll replace the name w_i with the name p_i in the weighted average expression.

$$average = \frac{\sum_{i=1}^{n} p_i x_i}{\sum_{i=1}^{n} p_i} \tag{2.13}$$

If the x_i are a complete set of possible random events (e.g. the integers two through twelve when rolling a pair of dice), the sum of all the probabilities must equal one. In this case, the summation in the denominator is one and it may be eliminated from the expression:

$$EV = \sum_{i=1}^{n} p_i x_i \tag{2.14}$$

This weighted average is given the name expected value (*EV*) of the random variable *x*. Remember that a weighted average an average of a list of numbers with each number weighted by how many times that number occurs when calculating the average. When the weighting list is a probability, we're generalizing "how many times this number occurs" to "what is the probability of this number occurring each time we try this."

As an example, let x_i represent the number of hours that a certain light bulb will last. The manufacturer has measured thousands of these bulbs and characterized their lifetime to the nearest 100 hours. Table 2.1 shows the probability of bulb lifetimes.

Just as a check – note that the sum of all the probabilities is 1.00. Any other result would mean that a mistake was made somewhere. Figure 2.2 is a graph (plot) of this PDF. Note that it is mentioned in the body of this figure that the original data has been rounded to the nearest 100 hours. It is important that this

Table 2.1 Lightbulb lifetime data distribution.

I	Lifetime	Probability
1	1	.02
2	2	.04
3	3	.10
4	4	.22
5	5	.30
6	6	.21
7	1	.07
8	2	.03
9	3	.01

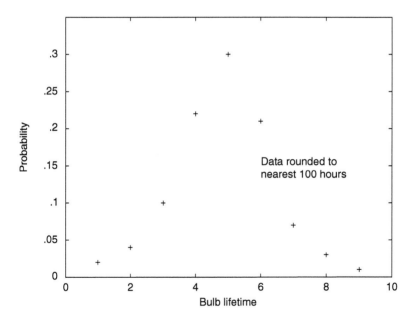

Figure 2.2 PDF of lightbulb lifetime data.

fact be mentioned somewhere in the discussion of bulb life statistics; otherwise erroneous conclusions about the resolution of the data might creep in.

Reviewing the subscripted variable notation, $x_4 = 1100$, $x_8 = 1500$, $p_4 = .22$, $p_9 = .01$, etc.

The *EV* of *x* is therefore

$$EV = \sum_{i=1}^{9} x_i p_i = (800)(.02) + (900)(.04) + (1000)(.10)$$
$$+ (1100)(.22) + (1200)(.30) + (1300)(.21) \qquad (2.15)$$
$$+ (1400)(.07) + (1500)(.03) + (1600)(.01)$$
$$= 1186$$

This means that, if you were to buy a few thousand of these bulbs, you would expect the average lifetime to be about 1186 hours. Since the original data was only reported to the nearest 100 hours, it would be correct to report this *EV* as 1200 hours.

The Basic Coin Flip Game

In Chapter 8, we will go into considerable detail analyzing the simple coin flip gambling game. This game was introduced in Chapter 1; it is a good example for studying *EV* calculations. The rules of the game are simple: You or your opponent tosses (flips) a coin. If the coin lands heads up you collect a dollar from your opponent; if it lands tails up you give you a dollar to your opponent. Assuming a fair coin, the probability of either occurrence is .5. Taking the winning of a dollar as a variable value of +1 and the losing of a dollar as the value −1, the *EV* is simply

$$EV = \sum_{i=1}^{2} x_i p_i = (+1)(.50) + (-1)(.50) = .50 - .50 = 0 \qquad (2.16)$$

This seems odd: if we flip a coin just once, I must either win a dollar or lose a dollar. There are simply no other choices. We seem to have a contradiction here, as was discussed in Chapter 1. In this game you might never actually see the *expected* value. The *EV* is telling us that if you flip the coin thousands of times, you should expect the difference between your winnings and my losses to be only a small percentage of the number of times a dollar changed hands (the number of flips). This is, again, what we mean by the Frequentist definition of probability.

PDF Symmetry

Figure 2.3 is an example of a symmetric PDF. A vertical line drawn through *x* = 3 is the axis of symmetry, three being the value of *x* from which things look the same both to the right and to the left (mirror images).

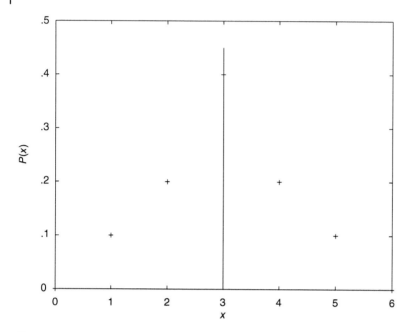

Figure 2.3 A symmetric PDF.

The *EV* of this PDF, calculated in detail, is:

$$E(x) = (.1)(1) + (.2)(2) + (.4)(3) + (.2)(4) + (.1)(5) = 3 \qquad (2.17)$$

The *EV* of *x* is identically (the *x* value of) the axis of symmetry. Is this a coincidence, or will this always be the case? We can explore this question by rewriting the *EV* calculation in a particular manner: for each value of *x*, write *x* in terms of its distance from the axis of symmetry. That is, for $x = 1$, write $x = 3 - 2$; for $x = 4$, write $x = 3 + 1$, etc. The *EV* calculation can then be rewritten as

$$E(x) = (.1)(3-2) + (.2)(3-1) + (.4)(3) + (.2)(3+1) + (.1)(3+2) = 3 \quad (2.18)$$

Rearranging the terms of Eq. (2.18),

$$E(x) = (.1)(3-2) + (.1)(3+2) + (.2)(3-1) + (.2)(3+1) + (.4)(3) = 3 \quad (2.19)$$

For the term containing $(3-2)$, there is a term with the same probability containing $(3+2)$. For the term containing $(3-1)$, there is a term with the same probability containing $(3+1)$. The symmetry property guarantees that there will be a balance of these terms for every term except the term $x = 3$ itself.

Now look at the term containing $(3-2)$ and the term containing $(3+2)$:

$$(.1)(3-2) + (.1)(3+2) = (.1)(3) - (.1)(2) + (.1)(3) + (.1)(2)$$
$$= (.1)(3) + (.1)(3) \qquad (2.20)$$

What has happened? The $-(.1)(2)$ and $+(.1)(2)$ pieces have canceled out, as if they had never been there in the first place. This operation can be repeated for every pair of terms present, leaving us with

$$E(x) = (.1)(3) + (.2)(3) + (.4)(3) + (.2)(3) + (.1)(3)$$
$$= (.1 + .2 + .4 + .2 + .1)(3) = (1)(3) = 3 \qquad (2.21)$$

Since we know that the probabilities must *always* add up to 1, we see that for a symmetric PDF the *EV* of *x* must always be the axis of symmetry itself.

This example of a symmetric PDF has an odd number of *x* values. What about a symmetric PDF with an even number of *x* values, such as the simple coin flip game? One of the problems at the end of this chapter will be to show that a symmetric PDF with an even number of *x* values cannot have an *x* data point at the axis of symmetry.

In the case of a symmetric PDF with an odd number of points, if the point on the axis of symmetry also is the point that has the largest probability, then the *EV* of *x* is also the most likely value of *x*.

Another special case is that of a PDF where the probabilities of all the variable values are the same. This may, but it doesn't have to, be a symmetric PDF.

Consider the *PDF* $x = [0, 1, 2, 4, 7], w = [.2, .2, .2, .2, .2]$.

If all of the probabilities are the same and they all must add up to 1, then these probability values must of course be = 1/(number of values). In this case, there are 5 random variable values, so each probability must be just 1/5 = .2. The *EV* of *x* is

$$E(x) = (.2)(0) + (.2)(1) + (.2)(2) + (.2)(4) + (.2)(7)$$

Rearranging this by "pulling out" the common factor of .2,

$$E(x) = (.2)(0 + 1 + 2 + 4 + 7) = \frac{0 + 1 + 2 + 4 + 7}{5} = 2.8$$

When all of the probabilities are equal, the *EV* is just the average value of the random variable values.

A word about averages: the average as calculated above is sometimes called the *arithmetic average*, to distinguish it from the *geometric average*, which isn't of interest here. It is also sometimes called the *arithmetic mean*. The mean must be no smaller than the smallest value in the group and no larger than the largest value in the group. Intuitively, it's a number that somehow "summarizes" a characteristic of the group. For example, if the mean age of a class of children is 7.2 years old, you would guess that this might be a 2nd grade class, but it is certainly not a high school class. Another indicator about the properties of a group of numbers is the *median*. The median is the number that has as many numbers to the right of it (on the graph) as it does to the left of it, i.e. bigger than it and less than it. In the case of a symmetric PDF with an odd number of terms, the median x value is identically the mean as well as the axis of symmetry. A third measurement is called the *mode*. The mode is the number in a list of numbers that is repeated the most. In the list of numbers used a few pages ago to explain indices and weighted averages, the mode is 5.

Standard Deviation

Suppose you are a manufacturer of glass marbles. You have a very tightly controlled process that manufactures one ounce (oz) marbles. You sell these marbles in three-pound (48 oz) bags, so there should be 48 marbles in each bag. For the past few months you've been having trouble with the machine that loads the marbles into the bags – it's gotten erratic. Just to be safe until you get the machine fixed, you decide to set your machine to put 50 marbles into each bag. This way no customer should ever get less than 48 marbles – but still, there are customer complaints.

The easiest way to track how many marbles are loaded into each bag is to count some of them. You grab 100 bags of marbles and count the number of marbles in each bag. Table 2.2 shows what you find.

No wonder some customers are complaining! While most of the customers are getting 48 or more marbles, some customers are getting fewer marbles than they've been promised!

Table 2.2 is easily converted into a PDF. Since there were 100 bags of marbles weighed, the probability of getting 50 marbles is just 35/100 = .35, etc. In other words, take the Nr of Bags column and divide all the entries by the total number of bags of marbles. A graph with Nr of Marbles in Bag as the x axis and Nr of Marbles as the y axis is the desired PDF. This PDF is shown in Figure 2.4.

To really know exactly what your marble factory is shipping, you would have to count the number of marbles in each and every bag of marbles that you produce. Since this is extremely time consuming, you would like to *sample* your production by only selecting a bag of marbles to count every now and then. The question is, what do I mean by "now and then" and how do you know when your sample population truly represents your entire distribution? One extreme is counting the marbles in every bag. Alternatively, you could count the marbles in a bag pulled off the shipping dock once a month. Or, you could grab a bag every hour until you have 100 bags and look at the distribution. While the first two choices are not very good (too time consuming or not representative of the true distribution), how do we know just how useful the information we're getting from the last choice really is? This is a discussion for another chapter. For now, assume that the 100-bag sample represents the factory production very well.

Table 2.2 Marble manufacturer's 100 bag production sample.

Nr of marbles in bag	Nr of bags
42	2
43	0
44	0
45	0
46	0
47	8
48	3
49	20
50	35
51	14
52	0
53	14
54	0
55	4

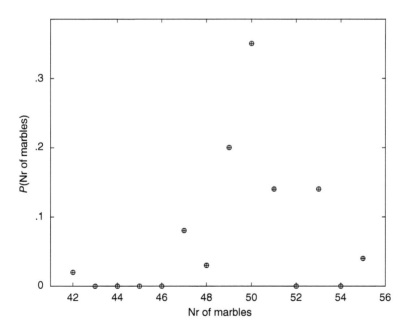

Figure 2.4 PDF of number of marbles in bags data.

Figure 2.4 shows that the bag filling machine is not doing a good enough job of counting marbles. Let's assume that you reworked the machine a bit, ran it for a while, repeated the 100-bag counting exercise and got Figure 2.5. This PDF indicates that fewer customers are now getting less than 48 marbles, but the situation still isn't acceptable. Let's refer to the PDFs of Figures 2.4 and 2.5 as *Poor* and *Marginal*, respectively, and tabulate them in Table 2.3.

Calculating the *EV* of both distributions, without showing the arithmetic here, we get 50.10 for the *Poor* distribution and 49.97 for the *Marginal* distribution. This seems odd. The goal is to minimize (ideally, to eliminate) bags with less than 48 marbles. It would seem as if raising the *EV* of the number of marbles in the bag would make things better – but in this case the distribution with the higher *EV* gives worse results (number of bags with less than 48 marbles)!

We need something more than the *EV* in order to characterize the performance of our machine. We need a way to quantify the difference between these two distributions in a meaningful way; there is something more going on here than the *EV* alone is telling us.

Table 2.4 is an expansion of Table 2.3; for each distribution we've added the number of marbles in the bag in terms of their difference from the *EV*.

The numbers in the two new columns in Table 2.4 are both positive and negative, indicating values higher than or lower than the *EV*(s), respectively.

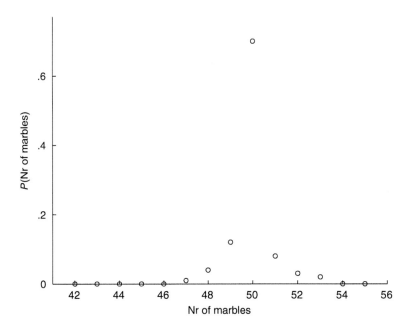

Figure 2.5 PDF of number of marbles in bags data, reworked bag filling machine.

Table 2.3 Marble manufacturer's *Poor* and *Marginal* PDFs.

Nr of marbles in bag	Poor PDF	Marginal PDF
42	.022	0
43	0	0
44	0	0
45	0	0
46	0	0
47	.08	.01
48	.033	.04
49	.20	.12
50	.35	.70
51	.14	.08
52	0	.03
53	.14	.02
54	0	0
55	.4	0

Table 2.4 Table 2.3 extended to include differences from expected values.

Nr of marbles in bag	Poor probability	Nr − *EV*	Marginal probability	Nr − *EV*
42	.022	42 − 50.10 = −8.10	0	42 − 49.97 = −7.97
43	0	43 − 50.10 = −7.10	0	−6.97
44	0	−6.10	0	−5.97
45	0	−5.10	0	−4.97
46	0	−4.10	0	−3.97
47	.08	−3.10	.01	−2.97
48	.033	−2.10	.04	−1.97
49	.20	−1.10	.12	−.97
50	.35	−.10	.70	+.03
51	.14	+.90	.08	+1.03
52	0	+1.90	.03	+2.03
53	.14	+2.90	.02	+3.03
54	0	+3.90	0	+4.03
55	.4	+4.90	0	+5.03

We aren't as much interested in whether these numbers are higher or lower than the $EV(s)$ as we are in how far away they are from their respective EVs. We can keep the information we're interested in while eliminating the information we're not interested in by simply squaring these numbers. Remember that the square of a number is always positive. Table 2.5 shows the results of squaring the numbers in columns 3 and 5 of Table 2.4.

Equation (2.14) gives us the EV of a list of random numbers, x_i. We can calculate the EV of any expression in the same way. In particular, for the Poor distribution calculation in Table 2.5,

$$\text{var} = \sum_{42}^{55} p_i (x_i - EV)^2 = (.022)(65.61) + (0)(50.41) + \cdots + (.4)(24.01)$$
$$= 4.71 \tag{2.22}$$

This expression, called the Variance, is the EV of the square of the differences between the variables and their mean.

The results of these calculations are 4.71 for the poor PDF and .75 for the marginal PDF.

Variance is a measure of how much we can expect results to *vary* from their EV with repeated trials. While the marginal PDF has the higher EV of these two PDFs, which we might think means that we're going to get many unhappy customers, the marginal PDF's variance is so much the lower of the two that this PDF would

Table 2.5 Table 2.4 rewritten to show square of differences from expected values.

Nr of marbles in bag	Poor probability	$(Nr - EV)^2$	Marginal probability	$(Nr - EV)^2$
42	.022	65.61	0	63.82
43	0	50.41	0	48.58
44	0	37.21	0	35.64
45	0	26.01	0	24.70
46	0	16.81	0	15.76
47	.08	9.61	.01	8.82
48	.033	4.41	.04	3.88
49	.20	1.21	.12	.94
50	.35	.01	.70	.00
51	.14	.81	.08	1.06
52	0	3.61	.03	4.12
53	.14	8.41	.02	9.18
54	0	15.21	0	16.24
55	.4	24.01	0	25.30

actually result in fewer customers getting fewer marbles than the package promises. Another way of putting it is that the lower the variance, the more tightly the numbers will be "bunched" about the *EV*. An interesting property of squaring the error terms is that since the squares of numbers climb much faster than the numbers do themselves ($2^2 = 4$, $3^2 = 9$, $4^2 = 16$, etc.), a single moderate deviation will have a much larger effect on the variance than will many small deviations.

The properties of the variance have made it almost the universal approach to characterizing how well a list of numbers is grouped about its *EV*. While we described the variance for a PDF, the same definition can be used for any list of numbers. Using the tools of calculus, the same definition can be extended to continuous PDFs.

A useful identity for the formula for the variance is

$$\text{var} = \sum p_i \left(x_i - E(x) \right)^2 = \sum p_i x_i^2 - 2E(x) \sum p_i x_i + \sum p_i \left[E(x) \right]^2$$
$$= E\left(x_i^2 \right) - 2\left[E(x) \right]^2 + \left[E(x) \right]^2 = E\left(x_i^2 \right) - \left[E(x) \right]^2 \tag{2.23}$$

We've introduced a notational shortcut above – When there are no upper or lower indices shown, it usually means "sum over the entire list." This is also sometimes written as $\sum_i$, which means "sum over all *i*."

An annoying factor when dealing with variances is that they do not have the same *units* as the underlying random variable. In the above example, we are dealing with numbers of marbles, but the variance is measured in number of marbles squared. In order to be able to picture what to expect, it is common to use the square-root of the variance, known as the *standard deviation*. In this example, the standard deviation for the poor PDF is 2.17, while the standard deviation for the marginal PDF is .87. Standard deviation is used so often in probability and statistics calculations that its usual symbol, σ (lower case Greek sigma), is universally recognized as standing for the standard deviation and often people will simply say "sigma" to mean "standard deviation."

Our next example is the simplest distribution to picture, the *uniform* distribution. A uniform distribution of *n* events is defined as a distribution in which each event is equally likely. The probability of each event is therefore simply $1/n$. (The coin flip PDF is an $n = 2$ uniform distribution in which each event has a probability of $1/n = \frac{1}{2}$.)

Figure 2.6 shows the PDF of an example of a uniform distribution of integers. From the figure you can see that this is a uniform PDF with a starting, or lowest, value of $x = 1$, and an ending, or highest, value of $x = 20$. The *EV* of this distribution is very easy to calculate because the probabilities are all the same, so they don't really need an index (the subscript *i*), and the probability can be written outside the summation sign:

$$EV = \sum_{i=1}^{n} p_i x_i = p \sum_{i=1}^{n} x_i = \frac{1}{n} \sum_{i=1}^{n} x_i \tag{2.24}$$

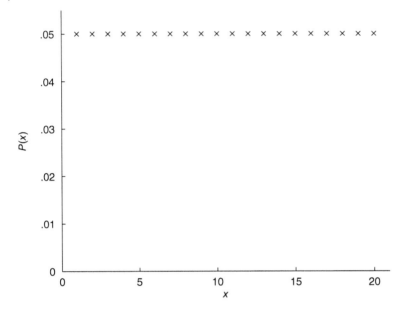

Figure 2.6 PDF of uniform distribution of integers.

Come to think of it, this is exactly the formula for the simple average of n items x_i, which is a property of uniform distributions that was described earlier in this chapter. To formally evaluate this expression, we would need to figure out how to express the locations of each of the x_i depending on the lowest value, the highest value, and the number of points. This isn't really that big a job, but fortunately it's an unnecessary job. Just look at the graph of this PDF – regardless of the lowest value, the highest value, and the number of points, this graph is symmetric about some value of x. Let's call the lowest x value x_{low} and the highest value x_{high}. We want to find the value of x that is equidistant from x_{low} and x_{high}, or in other words, the midpoint of the line between them. We know that this point will be the EV of x for a symmetric PDF.

$$Expected\ Value = \frac{x_{low} + x_{high}}{2}$$

For the PDF shown in Figure 2.6, the EV of x is therefore 11.5.

Next, consider some standard deviations of small uniform distributions. For a three-point distribution between $x = 1$ and $x = 3$ (Table 2.6), the EV is 2 and the standard deviation is .816.

Move this same distribution over so that it is a 3-point uniform distribution between $x = 0$ and $x = 2$, as shown in Table 2.7.

Comparing the above two distributions, we see that the EVs are different. That this should happen is clear from the symmetry argument – the EV has to

Table 2.6 Standard deviation calculation for three-point uniform distribution, $1 \leq x \leq 3$.

i	x_i	P_i	$x_i p_i$	$(x_i - EV)^2$	$P_i(x_i - EV)^2$
1	1	.333	.333	1	.333
2	2	.333	.667	0	0
3	3	.333	1.000	1	.333
			$EV = 2$		Var = .667
					$\sigma = .816$

Table 2.7 Standard deviation calculation for three-point uniform distribution, $0 \leq x \leq 2$.

i	x_i	P_i	$x_i p_i$	$(x_i - EV)^2$	$P_i(x_i - EV)^2$
1	0	.333	0	1	.333
2	1	.333	.333	0	0
3	2	.333	.667	1	.333
			$EV = 1$		Var = .667
					$\sigma = .816$

Table 2.8 Standard deviations of several three-point uniform distributions, different widths.

Width	σ
1	.408
2	.816
4	1.63

be in the "middle" of the distribution. The standard deviation, on the other hand, is the same for both distributions. This is because standard deviation is a property of the shape of the distribution, not its location on the x-axis.

Now look at two other cases. Narrow the distribution so that it only extends from 0 to 1 and widen it so that it extends from 0 to 4. The calculations parallel those above, so only the results are shown in Table 2.8.

For a three-point uniform distribution of width $X_{high} - X_{low}$, it appears that the standard deviation (σ) scales directly with the width of the distribution. In other words, the standard deviation is a measure of how much the points (in this example three points) are spread out about the EV. If all three points lied on top of each other, that is they all had the same value of x, then the standard deviation would be identically 0.

Finally, we'll keep the width of the distribution constant but vary the number of points. For distribution between 1 and 3, Table 2.9 shows σ for several numbers of points (n).

While the EV of x does not depend on the number of points in the uniform PDF, the standard deviation clearly does depend on the number of points. Figure 2.7 shows how the standard deviation varies with the number of points.[2] A calculation that is outside of the scope of this book shows that, as n gets very large (approaches infinity), the standard deviation approaches .5774. At $n = 100$, the standard deviation is .5832, which is only about 1% larger than the

Table 2.9 Standard deviations of several three-point uniform distributions, width = 2, different numbers of points.

n	σ
2	1.000
3	.816
4	.745

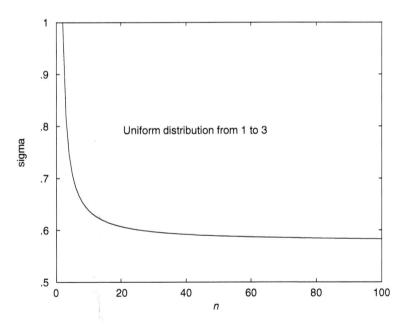

Figure 2.7 Standard deviation of uniform distribution of integers vs. number of points.

---

2 Figure 2.7 is one of the figures where the curve looks continuous because there are many tightly spaced points. It is, however, a discrete set of points, one for each value of n.

standard deviation for an infinitely large number of points. In general, for n very large, the standard deviation of a uniform distribution is

$$\sigma = \frac{x_{high} - x_{low}}{\sqrt{12}} = \frac{x_{high} - x_{low}}{3.464} \tag{2.25}$$

Cumulative Distribution Function

We've shown how σ describes the "width" of a distribution about its *EV*. Now let's quantify this. First we will describe the Cumulative Distribution Function (CDF).

Return to the uniform PDF between $x = 1$ and $x = 3$. Consider a distribution with 21 points (Figures 2.8 and 2.9). The cumulative distribution is plotted by starting (with a value of 0) at an x value lower than any x value in the PDF and then stepping to the right (increasing x) at the same x step sizes as the PDF and plotting the sum of all the probabilities that have been encountered so far.

This isn't as difficult as it sounds. For $x < 1$ (x less than 1), the value of the CDF is 0. At $x = 1$, the PDF value is .048, and the CDF value steps to .048. For $x = 1.1$, the PDF value is .048, so the CDF value steps to .048 + .048 = .096. For $x = .2$, the PDF value is .048, so the CDF value steps .096 + .048 = .144, etc.

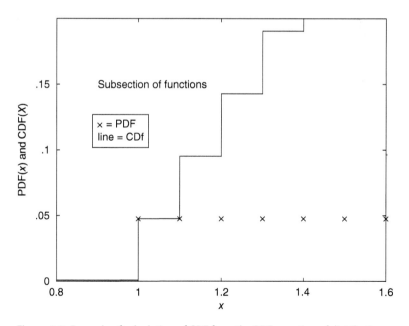

Figure 2.8 Example of calculation of CDF from the PDF – section of distribution.

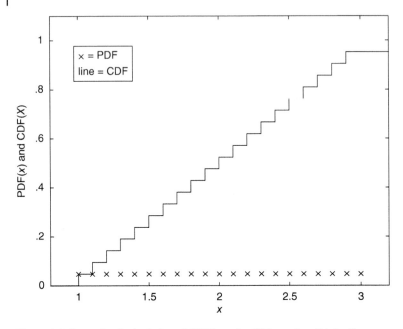

Figure 2.9 Example of calculation of CDF from the PDF – entire distribution.

Figure 2.10 shows the PDF and CDF for a continuous distribution function. This function extends to infinity in both $-x$ and $+x$ directions, so the CDF is >0 and <1 for all x. Since the PDF gets very small for very small $-x$ and very large $+x$, the CDF gets very close to 0 and 1 in these extreme regions of x.

We can immediately write down 3 important properties of the CDF:

1) If the PDF has a value of x below which the probability is 0, then the CDF is 0 below this same value of x. For a discrete PDF with a finite number of points, this will always happen.
2) As x increases, the CDF can either increase or remain constant, but it can never decrease. This property comes straight out of the rule above for generating a CDF – the word *"subtract"* is never mentioned.
3) If the PDF has a value of x above which the probability is 0 (there are no more data points), then the CDF is 1 at and above this same value of x. This is because, if we're above the value of x where all nonzero probabilities have been shown, then we must have covered every possible case and we know that all of the probabilities in a PDF must add up to 1.

In Figure 2.9, we see a uniform distribution of 21 points between $x = 1$ and $x = 3$. In Figure 2.7, we see that this distribution has a σ of approximately .6. Also, the *EV* for this distribution is 2.0.

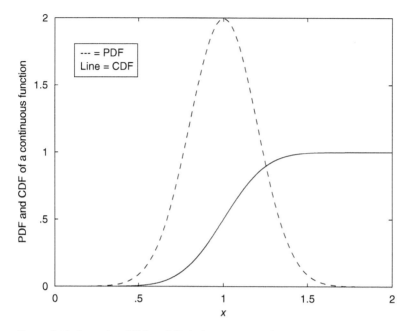

Figure 2.10 Example of PDF and CDF of a continuous function.

Adding the mean and sigma gives us 2.6; subtracting sigma from the mean gives 1.4.

In Figure 2.11, there is a box whose left and right sides lie on 1.4 and 2.6, respectively. All of the points inside or on the box are the points that lie within 1σ of the EV of x. There are 13 points inside the box out of the total of 21 points of the PDF. This means that 13 out of 21, or 62% of the points lie within 1σ of the EV of x.

The Confidence Interval

If we had a 21-sided die with the faces marked with the 21 values of this PDF and we rolled this die many times, we would expect the results to fall within 1σ of the EV of x 62% of the time. Since this is a symmetric PDF, we would also expect to fall below this range 19% of the time and above this range 19% of the time. For this distribution, we say that our 62% confidence interval is $\pm 1\sigma$ wide.

The confidence interval is a measure of how well we know a number. Knowing σ alone isn't enough because the confidence interval depends both on σ and the shape of the specific PDF being dealt with.

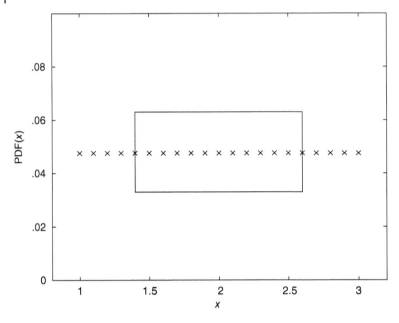

Figure 2.11 Discrete uniform distribution showing standard deviation region.

Final Points

1) In the CDF discussion above, a 21-point PDF was used as an example. In the CDF, the steps in x were .10 and the steps in y were about .05. We can talk about a region 1.00 wide, or 1.05 wide, or 1.10 wide, etc.; but we have to think about whether or not it makes sense to talk about a region, say, 1.12 wide. At least one of the end points of this region would fall in between X values and not be well defined. If this PDF really came about from a 21-faced die, then the integer X points are all there and we simply have to be careful what regions we try to talk about.

On the other hand, suppose we are weighing, say, fish from a commercial catch using a scale that only gives us weights to the nearest pound. We will end up with data that can be graphed as a PDF, but we know that the actual fish weights mostly fall between the points on the graph. In this example, our *measurement resolution* is 1 pound. Our "fish PDF" is a discrete PDF approximation to what's really a continuous PDF. We should treat this mathematically by drawing a continuous PDF that – in some sense – is a *best fit* to the discrete data. This is a serious topic for an in-depth probability study that is beyond our scope. The next chapter will show how many practical cases turn out.

2) The definition of standard deviation given above gives an answer (zero) for a distribution that consists of only one point. This might not seem to be more than a nuisance because nobody would ever bother to calculate the standard deviation for one point, much less ponder what it might mean. However, it does reveal an underlying issue. Since we need at least two points to meaningfully discuss deviations of any sort, we really should base our standard deviation calculations on one less x variable point than we actually have. This is referred to this as a "number of degrees of freedom" issue – we have one less degree of freedom than we have x variable points.

   In order to correct for this we go back to the definition of variance. The variance, if you recall, is the weighted average of the squares of the difference between the x points and the *EV*. Since an average is basically a sum divided by the number of points being summed, n, we should replace n by $n-1$ in the calculation. One way of doing this is to multiply the variance as defined previously by the term $\frac{n}{n-1}$. Since we're usually dealing with standard deviations rather than variances, we can apply this correction directly to the standard deviation by multiplying the standard deviation as defined by the square root of the above correction, i.e. by $\sqrt{\frac{n}{n-1}}$.

   If you've been trying to recreate the calculations in this chapter using functions in a spreadsheet or a pocket calculator and are wondering why the standard deviations didn't agree, this correction is the answer to your problem.

   Note that as n gets larger, this correction gets less and less important. For a small data set, such as $n = 3$, the correction is about 23%. For $n = 100$, the correction is about .5% and for $n = 1000$ it's about .05%. This is because $n/n-1$ gets very close to 1 as n gets large, so multiplying by it gets less and less significant.

Rehash and Histograms

In this chapter, we discussed lists of integers, lists of real numbers, weighted lists, discrete PDFs, continuous PDFs, CDFs, etc. While each of these items has a unique definition and purpose, often students take a while to *internalize* the concepts and subtleties. The purpose of this section is to give examples which compare and contrast these different items and hopefully make them all clearer, and to introduce histograms.

Suppose you are given a box containing a dozen small bags of hard candies. On the outside these bags look identical. When you pull a bag out of the box, you are choosing randomly.

You pull the bags out, one at a time, open them and count the number of candies in each bag. Table 2.10 shows what you find.

Table 2.10 List of number of candies in bags.

Bag nr	Nr of candies
1	17
2	14
3	16
4	16
5	14
6	15
7	14
8	20
9	20
10	18
11	14
12	21

In Table 2.10, we have assigned a Bag Number to each bag. This is an arbitrary convenience. It lets us refer to "the first bag, the fourth bag, etc." If we were to close up the bags, put the bag in the box, and repeat the process, the Bag Numbers would not repeat – the first bag most likely would not have 17 candies, etc. Another way of putting it is that this is an Unordered List.

This is a finite sized list, i.e. there are a finite number (in this case, 12) of entries. Also, this is a list of integers (aka counting numbers). This is important. The integer 14, for example, is an exact quantity. There cannot be a bag that has "slightly more than 14, but not quite 15, candies." A consequence of this is that the four bags that have 14 candies in them are fungible. We have no way of distinguishing any of these four bags from each other.

This latter fact lets us simplify the list by creating a Weighting List. The weighting list summarizes the "repeats" for us. Table 2.11 conveys the same information as Table 2.10, but in a more concise fashion. The Bag Nr column of Table 2.10 has been eliminated, but it was never of any use anyway. For convenience, we have ordered the list in ascending order of Number of Candies In a Bag (column one in Table 2.11).

Table 2.11 contains a row for 19 candies in a Bag, even though none of the bags contained 19 candies. Including this row or not is an arbitrary choice; it might help with a reader's perspective, it does no harm.

Table 2.11 shows how the candies are distributed in the bags, it is a Distribution Function. We could create a plot of this distribution function using the first column as the x-axis and the second column as the y-axis.

Table 2.11 Distribution of number of candies in bags.

Nr of candies in a bag	Nr of bags
14	4
15	1
16	2
17	1
18	1
19	0
20	2
21	1

Table 2.12 Repeat of Table 2.11 written as a PDF.

Nr of candies in a bag	Probability
14	.333
15	.083
16	.167
17	.083
18	.083
19	0
20	.167
21	.083

The total of the (entries in the) second column is 12. This is the number of items in the original list. Any other total would be an error.

We can convert Table 2.11 to a PDF by dividing each entry in the second column by the total number of bags (12), as shown in Table 2.12. This table tells us, for example, that the probability of a bag containing 15 candies is .083.

Using the data in Table 2.12, we can easily calculate the EV and the standard deviation (σ). Without showing the calculations, we get $EV = 16.6$ and $\sigma = 2.66$.

Table 2.13 shows the daily average temperatures taken for twelve days outside a home. Again, the Day Number column is not of mathematical significance except that it shows that this list is ordered. We may not use the ability to examine temperature trends (progress over time) but the information is here.

The temperatures in Table 2.13 are real numbers, not integers. This means that all real numbers, e.g. 71.4, 200.3, −99.22, and 1255 are legitimate entries.

Table 2.13 List of average daily temperatures.

Day nr	Temperature (°F)
1	62.4
2	71.3
3	80.0
4	59.6
5	71.3
6	58.6
7	69.4
8	71.3
9	66.5
10	81.2
11	74.1
12	77.7

They may be unrealistic temperatures, but they are still, formally, legitimate table entries.

The temperature 71.3 appears three time in Table 2.13. Can we proceed to develop a weights column as we did with the bags-of-candies example and create a PDF? Mathematically we could proceed and then calculate correct *EV* and σ values. However, whereas the numbers of candies were exact numbers representing fungible (undistinguishable) candy bags, the temperature measurements are approximations. How precise is our thermometer, how carefully did we read it? Is one of the numbers actually 71.305, another 71.29, and the third 71.300?

Therefore, in practice we are better off not simply *proclaiming* that a few entries that look identical are really identical. Instead, we formally gather similar entries into bins and create histograms.

Using Table 2.13 as our example, we first sort the list of temperatures in ascending order. Note that in doing this we are discarding the Day Nr column. Table 2.14 shows the sorted temperature list.

The (12) temperature entries listed vary from a low of 57.5 to a high of 82.5. We will choose to locate these entries in bins that are 5° wide. For convenience, we start one of the bins at 80.0°; the rest of the bin start and stop points then fall into place.

We can now plot a histogram of the temperature distributions (*y* axis) and the bin locations (*x*) axis. We could plot this as a simple graph plot, but it is typical to actually draw the full rectangles. If we divide the entries in the Nr-in-Bin column by the total number of temperature samples (12), we now have a PDF. Note that this is not a PDF of the original data, it is a PDF of our "eye

Table 2.14 List of average daily temperatures binned for a histogram.

Temperature	Bin range	Bin center (X axis)	Nr in bin (Y axis)
58.6	55–59.999	57.5	2
59.6			
62.4	60–64.999	62.5	1
66.5	65–69.999	67.5	2
69.4			
71.3	70–74.999	72.5	4
71.3			
71.3			
74.1			
77.7	75–79.999	77.5	1
80.0	80–84.999	82.5	2
81.2			

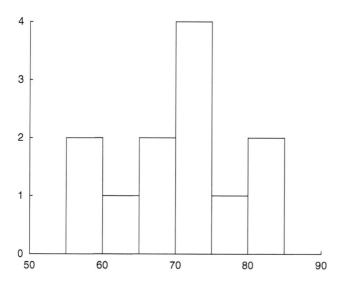

Figure 2.12 Histogram of temperature distribution data.

pleasing" approximation of the original data. It is not unique, we could have chosen bins 6 units wide, or 4 units wide, or ... .

Figure 2.12 shows this histogram. In this figure, the vertical axis shows the number of data points in each rectangle. We could equally well have shown this as the probability of a point being in each rectangle by scaling the vertical axis.

Remember that a histogram is just to help our brains *picture* the data. The true *EV* and standard deviation should be calculated from the original data, Table 2.13.

In practice, most data sets are generated from measurements of "something going on." These data sets will have a finite number of entries. However, it is possible for a discrete data set (integer or real numbers) to have an infinite number of entries and still be a valid PDF.

As an example, the infinite series of numbers (here used as probabilities) $\frac{1}{8} + \frac{1}{16} + \frac{1}{32} + \frac{1}{64} + \frac{1}{128} + \cdots$ can be shown to add up to exactly ¼. Table 2.15 shows two of these series plus the value $\frac{1}{2}$ lined up with integers, centered at $X = 0$. The third column shows the CDF. This list is terminated at $X_i = \pm 8$ so the sums are not complete, but the pattern and trend may be seen.

In all of the PDF examples thus far, there have been specific values for every X_i. We can also define a continuous PDF, where there are no specific values of

Table 2.15 A discrete PDF and CDF with an infinite number of entries.

X_i	Probability	CDF
...	≈0	.00000
−8	1/1024	.00098
−7	1/512	.00293
−6	1/256	.00684
−5	1/128	.04165
−4	1/64	.03027
−3	1/32	.06152
−2	1/16	.12402
−1	1/8	.24902
0	1/2	.74902
1	1/8	.87402
2	1/16	.93652
3	1/32	.96777
4	1/64	.98340
5	1/128	.99121
6	1/256	.99512
7	1/512	.99707
8	1/1024	.99805
...	≈0	1.0000

X_i, but rather a function defining possible values of X_i over some specific range or ranges. In this case, we replace the requirement that the sum of all the $X_i = 1$ with the requirement that the area under the curve of the function $= 1$.

In general, we need the calculus to calculate the area under a curve. However, there are many situations where simple geometry is all we need. The PDF shown in Figure 2.13 is an example of such a case. This PDF is described by the function

$$f(x) = \begin{cases} 0 & x < -10 \\ .1 + .01x & -10 \le x < 0 \\ .1 - .01x & 0 \le x \le 10 \\ 0 & 10 < x \end{cases} \qquad (2.26)$$

x is allowed to take on any real value. Since there are an infinite choice of x values, we cannot meaningfully talk about a probability for any particular value. Rather, we talk about the probability of finding x within a given range of values. For example,

$$P(-10 \le x \le 10) = 1$$

This range encompasses all of the nonzero range of $f(x)$. Therefore, it is the certain event.

$$P(0 \le x \le 10) = .5$$
$$P(-10 \le x \le 0) = .5$$
$$P(5 \le x \le 10) = .125$$

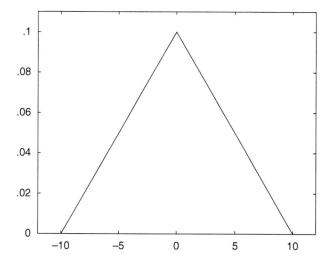

Figure 2.13 Example of continuous distribution where areas can be calculated without calculus.

$$P(-10 \le x \le 5) = .875$$

$$P(5 \le x \le 100) = .125$$

Continuous PDFs can be made up of several curves spanning disparate ranges with zero probability between them. Nothing changes, e.g. the total area under all the curves must = 1. The discussions of (the implications of) discontinuities in a curve, both a finite and an infinite number, get quite sophisticated. They are beyond the scope of this book but are an important part of modern probability theory.

Problems

2.1 For the function given below, evaluate the function for all values of the variable x in the list $x = [0, 7, 100, 121.3, -50, -50.2]$.

Function : $f(x) = 3x + 2$

2.2 Use the same list for x as in Problem 2.1, $x = [0, 7, 100, 121.3, -50, -50.2]$. x represents the number of jelly beans in a jar. Each jelly bean weighs 1.3 g. The jar weighs 100 g. What does the jar of beans weigh?

2.3 In this problem, we review *functional notation* and start using it exclusively. We have a generic name for a function, say "f" that we stipulate is a function of the variable x by writing "$f(x)$." This does *not* mean multiplication in any way, it just looks similar. We still must explicitly give the rule for the function f. In some cases the name of the function tells us the rule, e.g. $\sin(x)$ is the function "the sine of the variable x."

The way to tell when we're talking about a function and when we're talking about multiplication is to see if there's a variable "f" (or whatever other letter(s) you see) or not. If there is such a variable, then we're multiplying the variable f by whatever follows; if not, it's a function.

Here is an example:

1 Define the function. $f(x) \equiv \frac{3x-2}{7}$. The $\equiv$ symbol means "is defined as." It is not always used; often the equals sign (=) is used.

2 The notation for supplying a number for x is, (let's say $x = 22$):
$f(22) = \frac{3(32)-2}{7} = 9.14$.

The choice of the letter f is arbitrary. When we are dealing with a probability (PDF) we will often use the letter P, e.g. $P(x)$. This is not a necessary choice, but it helps to remind us of the nature of the function being described.

Using the same values of x as in Problems 2.1 and 2.2, $x = [0, 7, 100, 121.3, -50, -50.2]$, evaluate the function

$$f(x) = \begin{cases} 0 & x \leq 0 \\ x & 0 \leq x \leq 10 \\ x+3 & 10 < x \end{cases}$$

2.4 Given the two functions:

$$f(x) = x^2 + 4$$

$$g(y) = y - 7$$

Find $f(0)$, $g(0)$, $g(12)$, $f(3) + g(3)$, $f(3)g(3)$, and $f(g(8))$.

2.5 Define the *length* of a list as the number of entries in the list.
Given the list of numbers $L = [0, -4, -12.6, -4, -4, 7, 18, 22.2, 11.6, 0]$, find the length, the sum, and the mean.

2.6 Let's extend the use of named lists. Using the same list L as above, $L = [0, -4, -12.6, -4, -4, 7, 18, 22.2, 11.6, 0]$.
Define L_i as the i_{th} member of L, e.g. $L_2 = -4$, $L_6 = 7$, L_{11} does not exist. Let $s = $ length (L), let $P = [.1 \ .1 \ .1 \ .3 \ .2 \ 0 \ .2 \ 0 \ 0 \ 0]$.
Find the following

A $\displaystyle\sum_{i=1}^{s} L_i$

B $\displaystyle\sum_{i=1}^{4} L_i$

C $\displaystyle\sum_{i=1,3,5}^{s} L_i$

D $\displaystyle\sum_{i=1}^{s} P_i L_i$

E $\displaystyle\sum_{i=1}^{3}\sum_{j=1}^{3} L_i L_j$

F $\displaystyle\sum_{i=1}^{3}\sum_{j=1}^{i} L_j$

2.7 For the lists L and P as defined above, first verify that P is a legitimate list of probabilities for the set L, then sketch the PDF of the random variable list L with probabilities P, then find the expected value and the standard deviation.

$$L = [0, -4, -12.6, -4, -4, 7, 18, 22.2, 11.6, 0, P = \left[.1\ .1\ .1\ .3\ .20\ .2\ 0\ 0\ 0 \right]$$

2.8 Table below shows the measured heights of male and female students in a high school.[3]

Height (″)	Male	Female
54	1	3
55	2	13
56	1	11
57	3	20
58	2	34
59	6	44
60	12	65
61	22	85
62	28	100
63	40	137
64	33	200
65	41	224
66	74	201
67	93	184
68	85	99
69	99	75
70	100	54
71	85	45
72	105	31
73	75	28
74	40	18
75	36	11
76	12	0

---

3 This is not real data from any source. The numbers were made up for this problem.

There are separate columns for male and female heights ranging from 54″ to 74″. For males and females, separately find the expected values of the heights, the standard deviations of the heights, and the 95% width confidence factors for the heights. Then combine the male and the female columns into one "student" column and repeat the above.

2.9 We have a box with inside walls 10″ horizontal and .5″ vertical apart. On the bottom of the box we have printed a ruler with 1″ separation markings. We toss a coin with diameter slightly less than .5″ into the box (the coin diameter is essentially .5″, we just don't want it to wedge as it falls to the bottom). What is the probability that the coin lands on one of the markers?

2.10 Repeat Problem 2.9 but change the width of the box from .5″ to 2.0″.

2.11 A standard pair of dice are rolled and the larger number of the two individual die results is considered the result. (For example, a 3 and a 5 result from the roll; 5 is the result.) If the two numbers are the same, this number is the result.
Find the PDF.

3

Building a Bell

When a new random variable is created by summing other random variables, whether or not they all have the same probability distribution functions (PDFs), something special happens.

Rolling dice is an excellent example of this phenomenon. We know that there are six possible results for one die – the integers from 1 to 6. Since all six possible results are equally likely, the PDF of the rolling of a die is a uniform distribution with a probability of $\frac{1}{6}$ for each possible event.

Let's examine in detail what happens when we add up the results of several dice rolls. We can roll several dice or roll one die several times.

For the roll of one die one time, the PDF is simple, as shown in Figure 3.1.

Now let's get a second die, identical to the first. Roll them both and consider the sum of the two individual results. Clearly, this sum must be a number between 2 and 12 (1 on both dice and 6 on both dice, respectively). Table 1.2 shows all the possible results.

There are – of course – 36 possible combinations of the two dice, each combination having a probability of 1/36. However, since the sum must be an integer between 2 and 12, i.e. there are only 11 possible results. Also, except for sums of 2 and 12, there are more than one way to get any result. Table 1.2 is organized to show the number of ways of getting to each sum.

Figure 3.2 is the PDF corresponding to Table 1.2, for the sum of two rolled dice. This PDF doesn't look at all like the PDF for one die. Actually, it's triangular.

Let's continue this exercise for a while, adding the results of summing more and more dice. We won't show all of the combinations because the tables get very large.

For 3 dice there are $6^3 = 216$ entries, with sums between 3 and 18. The shape of the distribution is still fairly triangular (Figure 3.3), but curves are appearing. From a distance, you might even say that this function is "bell-like."

Probably Not: Future Prediction Using Probability and Statistical Inference, Second Edition. Lawrence N. Dworsky.
© 2019 John Wiley & Sons, Inc. Published 2019 by John Wiley & Sons, Inc.
Companion website: www.wiley.com/go/probablynot2e

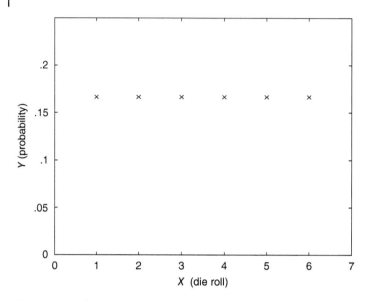

Figure 3.1 Uniform discrete PDF.

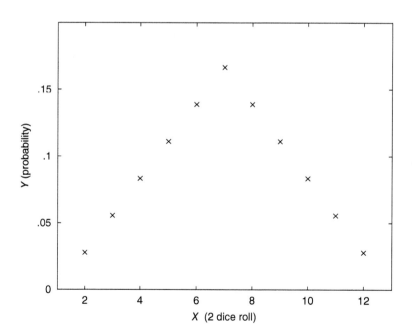

Figure 3.2 Sum of two of the PDFs of Figure 3.1.

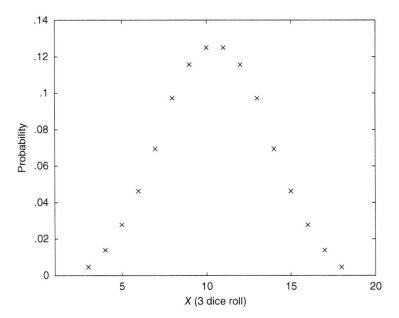

Figure 3.3 Sum of three of the PDFs of Figure 3.1.

Continuing, for the sum of 5 rolled dice we get a PDF that very definitely looks like a bell (Figure 3.4).

Without showing more examples, we'll assert that the sum of many rolls of a die produces a definitely bell-shaped PDF.

Since the PDF for one rolled die is simply a uniform distribution of integers (between 1 and 6), it's an easy leap to see that the sum of many uniform PDFs, whatever their upper and lower limits are, will always produce a bell-shaped curve as its PDF. Only the location of the peak (center of the curve) and the width of the curve will depend on the actual upper and lower limits of the original PDFs.

Table 1.2 illustrates the general concept of adding two discrete distribution functions. Each element of the first distribution is multiplied by every element of the second distribution and the results organized as in this table.

At this point it merely looks like an interesting curiosity has been demonstrated; the sum of many uniform PDFs produces a bell-shaped PDF. Actually, however, several other things have been shown. Remember that the sum of just two uniform PDFs produces a triangular PDF. This means that the PDF for, say, the sum of ten uniform PDFs is also the PDF for the sum of five triangular PDFs. In other words, the sum of many triangular PDFs also produces a bell-shaped PDF. Following this argument further, the sum of any combination of many uniform PDFs and many triangular PDFs produces a bell-shaped PDF. This curiosity is getting intriguing. Let's try another example.

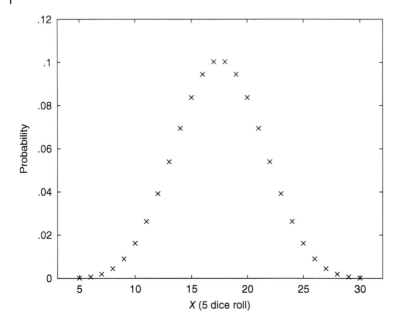

Figure 3.4 Sum of five of the PDFs of Figure 3.1.

This time we'll deliberately start with a distribution that's very "un-bell-like." The distribution shown in Figure 3.5 is about as un-bell-like as we can get. It looks like a "V."[1] We've even thrown a zero probability point into the distribution. If anything, this is closer to an upside-down bell than it is to a bell. Also, it's asymmetric. How can adding up asymmetric PDFs lead to a symmetric PDF?

Figure 3.6 is the PDF for two of these "lopsided V distributions" added together. This still doesn't look very bell-like, but something suspicious is beginning to happen. For one thing, the zero probability point in the original distribution has disappeared.

Figure 3.7 is the PDF for four distributions added together. This is not exactly a nice smooth bell curve, but the features are beginning to appear. Figure 3.8 is the PDF for ten of these distributions added together. At this point there is no doubt what's happening – we're building another symmetric bell.

It takes many summations before the lopsided-V PDF turned into a bell-shaped curve. This is because we started out so very "un" bell-shaped. Combinations of lopsided-V PDFs and uniform and/or triangular PDFs will look very bell-shaped much sooner.

---

1 This distribution has not been normalized to make in a *proper* PDF. Normalization would not change any of the discussion; we just must remember that if we want to look at probabilities – or confidence intervals – we must normalize the distribution function(s) that we're looking at.

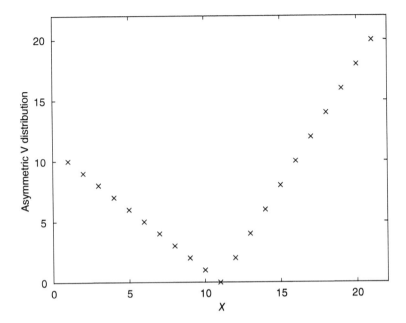

Figure 3.5 A very "Un-Bell-Like" discrete distribution.

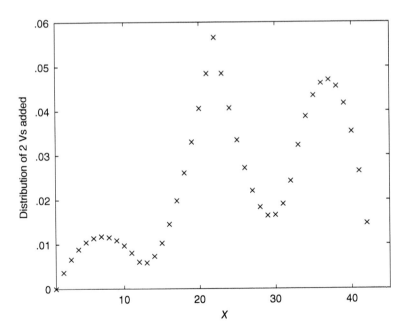

Figure 3.6 Sum of two of the PDFs of Figure 3.5.

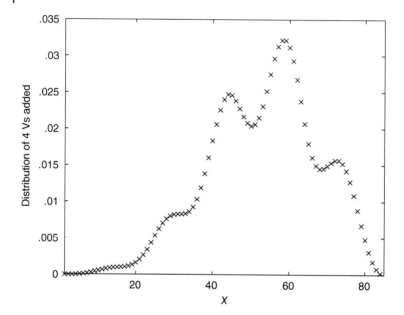

Figure 3.7 Sum of four of the PDFs of Figure 3.5.

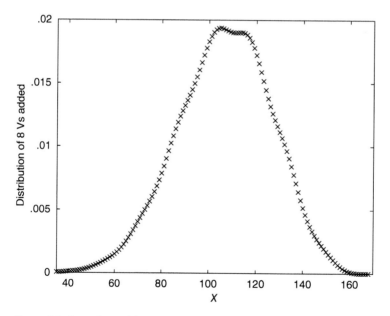

Figure 3.8 Sum of ten of the PDFs of Figure 3.5.

By now it should be clear that we're not looking at a coincidence. We're looking at one of the most intriguing mathematical relationships known. The formal name for this relationship is the Central Limit Theorem. Without getting into the precise mathematical language, this theorem says that the sums of the distributions of any PDF or combinations of any PDFs will, as we take more and more of these sums, approach a bell-shaped curve which is known as a normal, or Gaussian, distribution.

The central limit theorem comes in two versions. The first is the "Fuzzy" central limit theorem: When a collection of data comes about from many small and unrelated random effects, the data are approximately normally distributed. This is telling us that normal distributions are everywhere: daily temperature swings, people's heights and weights, your average driving time to work when you start at about the same time every morning, students' grades on exams. This is the (in)famous Bell Curve that is so often quoted, invoked, praised, and damned.

The actual Central Limit Theorem is a bit more involved. Look back at the light bulb manufacturer – he's the guy who takes a bunch of light bulbs from his factory every day and runs them until they burn out. He averages the lifetime of this *sample* of his production and writes down the result. He repeats this process every day and then looks at the distribution of these averages, which we'll call the *sample mean*. The Central Limit Theorem says that, as the size of these samples get larger and larger, this distribution approaches a Normal, or Gaussian, distribution. It also talks about the mean and standard deviation of the sample distribution and how they relate to the full population (of light bulb lifetimes), but we won't get into this in this chapter.

Since the normal distribution is so ubiquitous, it is important and we need to study it in some detail. The normal distribution is symmetric about its peak. This means that the mean equals the median, which equals the mode.

The full formula for the normal PDF is

$$Normal \ \ Distribution = \frac{1}{2\pi\sigma}e^{-\frac{(x-\mu)^2}{2\sigma^2}} \tag{3.1}$$

where μ is the mean and σ is the standard deviation.

This formula involves the exponent of e, which is the *base of natural logarithms*. This is a topic out of the calculus that isn't necessary for us to cope with, but we show the full formula for completeness. All scientific calculators, spreadsheets, and scientific programming languages will have the normal distribution available as a built-in function. The individual use instructions will vary, so it's impossible to quote them here.

The normal distribution as defined in Eq. (3.1) is a continuous distribution. However, we can use it comfortably for discrete distribution problems so long as there are enough (*x*) points for us to clearly recognize what we've got. Real situations such as our light bulb lifetime data are discrete and it's actually the continuous distribution that is the approximation to the real situation, not the other way around.

The continuous normal distribution extends (in *x*) from –infinity to +infinity. There is therefore an issue when we say, for example, that some students' grades for a given semester "fit nicely onto a bell-shaped curve." How is it possible for there to be negative values of *x* allowed? How is it possible for there to be positive values of *x* so large that some student might have scored better than perfect in her exams?

This dilemma is resolved by the term "approximate." The answer to this seeming contradiction lies in the fact that once you're more than several standard deviations away from the mean in a normal distribution, the probability of an event is so low that you can (almost always) ignore it. The caveat here is that oversight is necessary. Sometimes the normal distribution is not correctly approximating reality and other distributions are needed. In Chapter 6, we'll present the Poisson distribution as an example of this. On the other hand, the (fuzzy) central limit theorem can be trusted – real situations arising from many small random distributions added together look normal and if you just step back for a moment and look at what you're doing before proceeding you won't go wrong.

Figure 3.9 is the PDF of a normal distribution with $\mu = 3$ and $\sigma = 1$. The vertical lines indicate μ, 1σ, 2σ, and 3σ above and below μ. Figure 3.10 is the Cumulative Distribution Function (CDF) of the same normal distribution, with the same vertical lines. At 2σ below μ, the value of the CDF is .0228. At 2σ above μ, the value of the CDF is .9772 which is $1 - .9772 = .0228$ down from 1.0. This means that the area outside these 2σ limits in the PDF are equal to $.0228 + .0228 = .0456$ of the total area of the PDF. Equivalently, the area inside these limits is equal to $1 - .0456 = .944$ of the total area of the PDF. This is close enough to $.095 = 95\%$ that $\pm 2\sigma$ about μ is commonly accepted as the 95% confidence interval for the normal distribution. The 95% confidence interval is so often used as a reasonable confidence interval for everyday situations that it's often referred to as The Confidence Interval.

The $\pm 3\sigma$ range corresponds to an approximately 99.7% confidence interval, and a $\pm 4\sigma$ range corresponds to a slightly more than 99.99% confidence interval. Once we're more than 4σ away from the mean, the PDF is effectively 0. In most situations we don't worry about anything this far out. Consider both sides of this statement carefully. If we're looking at the distribution of weights of ears of corn or variations in annual rainfall or things like that, then we won't worry about anything outside the 99.99% confidence interval; we'll usually not worry about anything outside the 95% confidence interval. On the other hand, if we're looking at airline safety or drug safety issues, we'll certainly examine *at least* the 99.99% confidence interval.

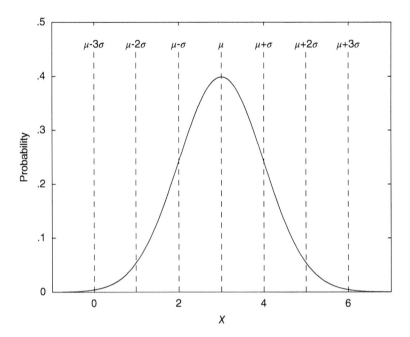

Figure 3.9 Normal PDF with $\mu = 3$ and $\sigma = 1$.

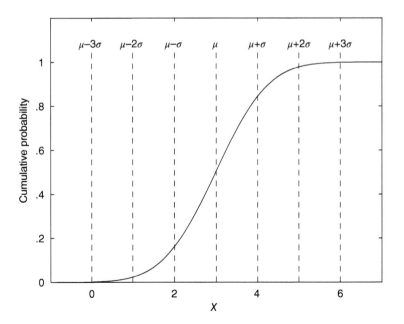

Figure 3.10 CDF of the normal PDF with $\mu = 3$ and $\sigma = 1$.

The $\pm 1\sigma$ interval (about the mean) corresponds to about a 68% confidence interval. Remember that this is equivalent to saying that this $\pm 1\sigma$ interval includes 68% of the points of a large sample of data that is normally distributed.

The CDF for a normal distribution is also called the Error Function (ERF).

As an example, let's take a look at the average of a term's test scores of a class of 50 students. The tests were graded on a 0–100% scale. Table 3.1 shows the list of scores, sorted in increasing order. We'll build a histogram of the distribution of these scores, so we can look at how they're distributed.

To create the histogram from a list of numbers, first we sort the list into increasing order. Next, we divide the x-axis into a number of equally size ranges, or bins. Then we create rectangles; for each rectangle the (bin) width is the range (they're all the same) and the height is the number of entries on the list that occur within that given rectangle.

Table 3.1 Sample test score data.

Score
39.1
39.4
40.8
41.3
41.5
41.9
42.4
42.8
43.6
44.6
45.1
45.8
46.1
46.3
46.5
46.7
47.2
47.4
47.4
48.0
48.1
48.1
48.2
48.2

Table 3.1 (Continued)

Score
48.4
48.4
48.5
48.8
49.1
49.6
49.6
49.8
50.1
50.7
51.2
52.7
53.4
53.8
54.3
54.5
55.5
56.0
56.4
56.7
58.3
58.7
59.6
59.9
61.0
61.9

Choosing the range size is art and experience as much as mathematics. If the range size is so small that only zero or one points on the list fit into any range, then we have recreated a graph of our list with boxes rather than points – this is useless. If the range size is so big that all the points on the list fit within it, then we have just drawn a box on a piece of paper (or computer screen) – this is also useless. Let us proceed by *trial and error*.

Figures 3.11–3.14 show the same data with the data divided into 4, 8, 12, and 16 bins, respectively. The number inside each rectangle is the number of data points that contributed to that rectangle.

At four bins (Figure 3.11), we can identify some gross characteristics of the distribution, but not much. At 16 bins (Figure 3.14), we are beginning to lose

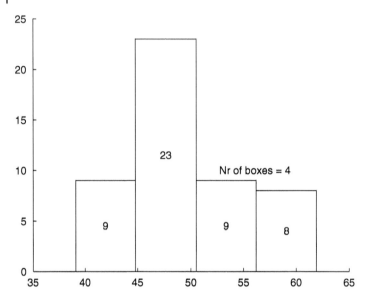

Figure 3.11 Histogram of data of Table 3.1, four bins.

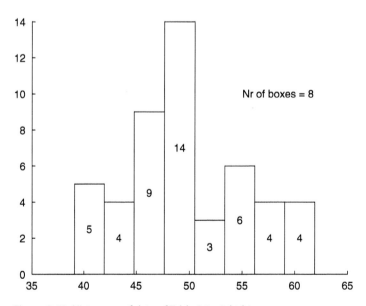

Figure 3.12 Histogram of data of Table 3.1, eight bins.

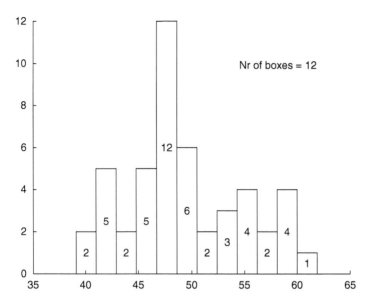

Figure 3.13 Histogram of data of Table 3.1, twelve bins.

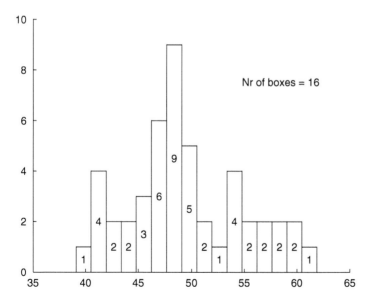

Figure 3.14 Histogram of data of Table 3.1, sixteen bins.

perspective because most of the bins have only one or two data points in them. Eight or twelve bins are the best choices for this data. Maybe 10 bins (not shown) would be the best compromise.

As was just demonstrated, there is a bit of "art" involved in generating useful histograms. Also, unfortunately, this is one of the places where the "lies, damned lies, and statistics" curse can ring true. By deliberately making the bins too large or too small, it's often possible to hide something that you don't want to show or conversely to emphasize some insignificant point that fits your agenda. You can't even claim that a person who does this is lying, because their truthful defense is that they did, indeed, use all of the data accurately when creating the histogram. The message here is that, when presented with a report where the conclusions will lead to or suppress some activity that you consider significant, get the data yourself and generate histograms using several different bin sizes so that you can draw your own conclusions about what the data is really telling you (or not telling you).

Table 3.1 is a list of test scores. It is not a distribution, discrete or otherwise. However, when we created the histograms we created distributions out of this data. If we normalized these distributions (by dividing the number of scores in each box by the total number of scores) then we would have PDFs.

We now have two questions to answer: what are the x_i values for these normalized scores, and which histogram is *correct*?

There is no conclusive answer to the latter question, especially when starting with a data set containing only 50 data points. The histogram to use is the one that "seems" to give us the best information about the nature of our data. However, once the histogram is chosen, the x_i values should be the x values of the (horizontal) centers of each of the boxes.

In practical situations, often we get some data points that are considered *outliers*. Outliers are a few points (few as compared to the total number of points) that don't make sense. For example, suppose we had one test score of 11. This wouldn't change the Expected Value, the Standard Deviation, or the shape of the histogram in any meaningful way. In some cases outliers are simply ignored. This, again, is dangerous territory in "deciding" which data points don't belong in our final report; we are injecting either a reasonable simplification or a lie, depending on your personal bias about what the data is telling us. Any data points that are not treated the same as is the main body of the data points need to be clearly explained.

Every list of numbers has an Expected Value and a Standard Deviation. For the test scores of Table 3.1, using the formulas from Chapter 2, these are

$$\mu = \frac{1}{50}\sum_{i=1}^{50}x_i = \frac{1}{50}(39.1+39.4+\cdots+61.9)=49.5 \tag{3.2}$$

$$\sigma = \sqrt{\frac{1}{50}\sum_{i=1}^{50}\left(x_i - \mu\right)^2} = \sqrt{\frac{1}{50}\left[\left(39.1-49.5\right)^2 + \cdots + \left(61.9-49.5\right)^2\right]} = 5.8 \quad (3.3)$$

Note that the weighting function (probability) in Eq. (3.3) is the probability of every number on the list, $\frac{1}{50}$. When the numbers in a list are real numbers, it is very unlikely that two or more of them will be identical, and we never develop a list of probabilities (weighting list). This is unlike the case when the numbers in a list are integers and exact repeats (identical entries) can occur.

When we create the bins, or ranges, for a histogram, we in effect take real numbers and group them into defined "you are identical" categories. Each bin has its own number of such "identical" entries and we get a weighting list consisting of the number of members of each bin. If we normalize this weighting function, we get a list of probabilities and we have a discrete distribution function. Remember that this distribution function is not unique; we chose the number of bins.

Figure 3.15 is a repeat of Figure 3.14, the 15-bin histogram, with a normal distribution having μ = 49.5 and σ = 5.8 as its parameters superimposed. The normal distribution is a good guess for a known formal distribution to approximate this data.

$\pm 2\sigma$ about the mean in a normal distribution should encompass 95% of the data. In a relatively small sample this is an approximation, but it should be a decent approximation. Now, $\mu - 2\sigma$ = 49.5 – 2(5.8) = 37.9 and

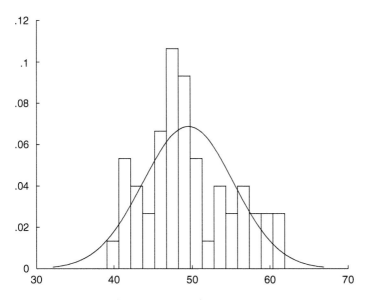

Figure 3.15 Repeat of Figure 3.14 with fitted normal distribution superimposed.

$\mu + 2\sigma = 49.5 + 2(5.8) = 61.1$. Looking at our list of numbers, this range encompasses everything but the highest number in the list (61.9). This is 49/50 = 98% of the data.

Actually, the list of numbers in Table 3.1 was generated using the Normal Distribution random number generator in a standard spreadsheet program, with a mean of 50 and a standard deviation of 5. The discrepancy is due to the small size of the data sample.

Returning to the class test scores, it's not uncommon for a teacher to use the statistics of test scores to determine grades. This is based on the belief that the average student should get the average grade and that test scores should fit pretty well onto a normal distribution (bell-shaped) curve. While these assertions might be reasonably true for a very large sample size, as we saw above they are only approximately true for a small group such as a class of 50. A typical procedure would be to give students whose grades are within $\pm 1\sigma$ of the mean a C, grades between σ and 2σ above the mean a B, grades more than 2σ above the mean an A, grades between σ and 2σ below the mean a D, and grades more than 2σ below the mean an F. In a large enough sample size, this would imply that there would be 68% of the students getting a C, 13.5% getting a B, 2.5% getting an A, 13.5% getting a D, and 2.5% getting an F.

In many situations, this procedure could give us utter nonsense. For example, suppose all of the grades are bunched between 49.0 and 51.0. We could still do the calculations and we could still allocate the grades, but the difference between the highest scoring student and the lowest scoring student is so small that they should probably all get the same grade – though it's hard to say what that grade should be. In any case, arguing that a few of the students in this group should get an A while a few of them should fail simply doesn't make much sense. In a situation like this you have to ask "is the difference between a grade of 49.0 and a grade of 51.0, in a group of this size, statistically significant?" We'll discuss just what this means and how to calculate it in the chapter on statistical inference.

Looking ahead to the discussion of statistical inference, we can expect that if we have a huge list of numbers whose expected value is 50 and whose sigma is 5, if we took many samples of 50, the sample means would average close to 50. As we took more and more samples, the sample mean would approach 50 and standard deviation of the sample means would get smaller and smaller. This is a qualitative way of saying that the more data we have, the better our knowledge of the background population is. We just haven't quantified this procedure yet.

In upcoming chapters, there will be situations where you'll find bell-shaped curves, e.g. how far a drunken individual taking a "random walk" should be expected to travel, or what to expect if you sit up all night tossing coins.

Problems

3.1 Consider these two lists of random variables. In both cases, both variables are equally probable

$$X_1 = \begin{bmatrix} +1 & -1 \end{bmatrix}$$

$$X_2 = \begin{bmatrix} +2 & 0 \end{bmatrix}$$

Build the PDF for multiple additions of each of these cases by repeatedly adding their PDFs, and then for repeatedly adding your results together. Either add the results of the previous addition to the original PDF or add the results of the previous addition to itself. Keep going until you are ready to make predictions about the PDFs after a large number of additions.

After each addition, calculate the Expected Value and the Standard Deviation of the latest PDF. Comment on the similarities and differences of the results of the two cases. Ultimately, does it matter which of these you work with?

Note: This can be done with pencil and paper for a few additions but working with a spreadsheet or a programming language will be necessary after a while.

3.2 In this problem the PDF is a normal distribution.
   A What is the CDF of $x = +$infinity for any μ and σ. No calculator needed.
   B Repeat the above for $x = 0$ in the case when $\mu = 0$, $\sigma = 3$.
   C When $\mu = 70$, $\sigma = 5$, what is the CDF at $x = 70$?
   D When $\mu = 70$, $\sigma = 5$ is the result of a large survey of men's heights (in inches). What percentage of men would you expect to be at least 70″ tall?
   E For the same distribution as above, what is the 95% confidence interval?

3.3 Before today's calculating tools were available, a book of values was necessary to find normal PDFs and CDFs. Since this book had no way of knowing what values of μ and σ you were using, the book only published one set of values, for $\mu = 0$, $\sigma = 1$. The PDF was referred to as the "standardized" normal distribution, often called $f(z)$, and the CDF was often called the "error function" or "standard error function."

The results of Problem 3.1 give enough information to show how to scale the values from such a book to anything needed. Pretend your only source of information is such a book (i.e. only use your calculator for $\mu = 0$, $\sigma = 1$ value calculations).

For a uniform distribution with $\mu = 70$, $\sigma = 5$ of men's heights, what percentage of men are between 66″ and 70″ tall?

3.4 Consider the (discrete) distribution data set shown in the below table. Scale the *y* values as necessary to make this a PDF, then find the expected value (EV) and standard deviation (σ).

X	Y
2	0
2.5	.2/3
3	.4/3
3.5	.6/3
4	.8/3
4.5	1/3
5	.4
5.5	.3
6	.2
6.5	.1
7	0

Calculate the CDF for the discrete data set. Then, for a normal distribution with mean = EV of the discrete set and standard deviation = that of the discrete set, calculate the CDF at the values of *x* given for the discrete set. In terms of confidence intervals, how good is this normal distribution at approximating the discrete set?

3.5 IQ test scores for a large group of students typically show normal distribution behavior. Suppose we have test scores for two groups:

Group A has $\mu = 100$, $\sigma = 30$
Group B has $\mu = 120$, $\sigma = 10$

On a world-wide basis, the average score is 100.
A What % of each group is above average?
B What is the % in each group with a score between 90 and 110?
C Suppose group A has 1000 people and group B has 400 people. If we put the test results for both groups into one big group, what would be the mean? Would the distribution still look normal?

3.6 We have 2 pairs of dice. One is a conventional pair, 6 faces on each die; the other has 4 faces on one die and 8 faces on the other die.
A For each die (3 different dice) what is the expected value and standard deviation?
B For each pair what is the expected value and standard deviation?

3.7 Repeat the last part of Problem 3.6, but for the differences rather than the sums of the die faces.

4

Random Walks

The One-Dimensional Random Walk

A random walk is a walk in which the direction and/or the size of the steps are randomly chosen. Typically, there are Probability Distribution Functions (PDFs) associated with the direction, size, and timing of the steps. In two or three dimensions the random walk is sometimes called Brownian motion, named after the botanist, Robert Brown. Brown described seemingly random motion of small particles in water.[1] In this chapter we'll tie together the idea of a random walk, which is another example of how the sum of experiments with a very simple PDF start looking normal, with the physical concept of *diffusion*, and a deeper look into what (frequentist) probability is.

Imagine that you're standing on a big ruler; it extends for miles in both directions. You are at position zero. In front of you are markings for +1 foot, +2 feet, +3 feet, etc. Behind you are markings for −1 foot, −2 feet, etc. Every ten seconds you flip a coin. If the coin lands on a head you take a step forward (in the + direction). If the coin lands on a tail you take a step backward (in the − direction). Each step moves you exactly one foot in your chosen direction. We are interested in where you could expect to be after many such coin flips.

If you were to keep a list of your coin flips, the list would look something like:

1) Head. Take 1 step forward
2) Tail. Take 1 step backward
3) Head. Take 1 step forward
4) Head. Take 1 step forward
5) Head. Take 1 step forward
6) Tail. Take 1 step backward, etc.

---

1 Brownian motion was analyzed by Albert Einstein and the success of his analysis became part of the proof of the existence of molecules and atoms as the building blocks for all matter.

Probably Not: Future Prediction Using Probability and Statistical Inference, Second Edition. Lawrence N. Dworsky.
© 2019 John Wiley & Sons, Inc. Published 2019 by John Wiley & Sons, Inc.
Companion website: www.wiley.com/go/probablynot2e

Now, assign a value of +1 to each step forward (heads) and a value of −1 to each step backward (tails). Your position after some number of coin flips would be just the algebraic sum of each of these values. For example, from the list above, you would have (+1 −1 +1 +1 +1 −1) = +2.

Figure 4.1 shows the results of a typical walk of this type. Random walks can have more complex rules. The size of each step can vary according to some PDF or algorithm, the time between steps can vary, and of course a real walk would be described in two dimensions with the direction of each step also being random. However, there is much to learn from the one-dimensional walk.

Three observations:

1) Since heads and tails are equally likely, the expected value of the sum is zero. In other words, the random walker shouldn't expect to get anywhere.
2) If you were gambling with the rules that you win $1 for each head and lose $1 for each tail rather than stepping forward and backward, this same procedure would tell you exactly your winnings (or losses) after a number of coin flips. This is mathematically the simple coin flip gambling game again.
3) Since we're looking at the sum of the results of a number of statistical events, we should suspect that a normal distribution will be involved.

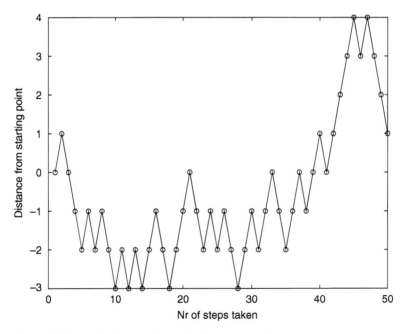

Figure 4.1 Example of a one-dimensional random walk.

After the first coin flip, you would have moved either 1 step forward or 1 step backward, with an equal (50%) probability of either. The PDF of this situation, Figure 4.2a, is quite simple. It is impossible to be at the starting point at this time. For that matter, after any odd number of coin flips it is impossible to be at the starting point. This, again, is directly analogous to the coin flip gambling game.

Continuing, Figure 4.2b shows the PDF for your position after two steps; it is just the sum of two of PDFs of Figure 4.2a. Figures 4.2c, d, e, and f are the PDFs after 3, 4, 5, and 6 steps, respectively.

It seems that we are once again *building a bell*. In this example, the mean (expected) value, μ, is identically 0 regardless of the number of steps.

Consider the probability of ending up where you started (at $x = 0$). Since this probability is identically 0 for an odd number of steps, we'll just present the even number cases. After n steps, the probability of being at $x = 0$ is shown in Table 4.1. The more steps you take, the less likely it is that you'll end up where you started ($x = 0$).

What about landing near $x = 0$, say at $x = 2$ or $x = -2$? The third column of Table 4.1 shows this.

While the probability of being at $x = 0$ starts at 1 and always decreases (with increasing n), the probability of being at $x = 2$ starts at 0, increases to a peak,

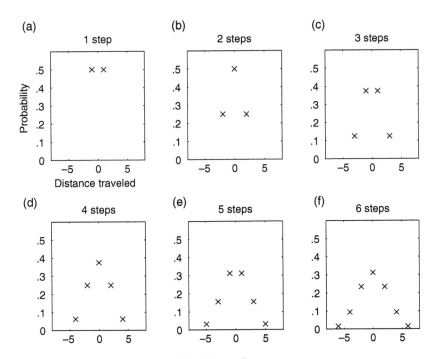

Figure 4.2 PDF of one-dimensional random walks.

Table 4.1 Various properties of a one-dimensional random walk.

Nr steps	$P(x = 0)$	$P(x = 2)$	Ratio	|Difference|	σ
0	1.000	0	0	1	
2	.375	.250	.667	.125	1.414
4	.313	.250	.800	.063	2.000
6	.273	.235	.861	.038	2.449
8	.246	.219	.890	.027	2.828
10	.226	.215	.951	.011	3.162
...	...	...	...	...	...
30	.144	.135	.968	.009	5.477

and then starts decreasing. This makes sense – before any steps it is impossible to be anywhere but at $x = 0$; after 2 steps it is likely to be at $x = 2$, after many steps there is a large opportunity to be in other places.

Comparing the probabilities of being at $x = 0$ and at $x = 2$, look both at the ratio of these probabilities and also their absolute difference. Columns 3 and 4 of Table 4.1 show these numbers.

Both of these latter calculations are telling us the same thing. The ratio of the two numbers is approaching 1 and the difference between the two numbers is approaching 0 as n gets larger. In other words, for very large values of n, we can expect the probability of finding yourself at $x = 0$ and at $x = 2$ to be almost the same – even though these probabilities are themselves both falling as n grows. This means that even though the expected value of x is always 0, the more steps you take the less likely it is that you'll be at $x = 0$ or anywhere near it.

As a last calculation, consider σ (uncorrected calculation), shown in the right hand column of Table 4.1 and also in Figure 4.3.

σ, it seems, gets larger as n gets larger, although its rate of growth with respect to n slows as n grows. In Figure 4.3, the numbers from Table 4.1 are superimposed over the curve $\sigma = \sqrt{n}$, a perfect fit.

Let us return to the discussion of coin flips. While we know that a large number of coin flips should yield an approximately equal number of heads and tails, we have not yet put our finger on in just what sense this is happening. By every measure so far, the more times we flip the coin, the less likely it is that we'll see an equal number, or even an approximately equal number, of heads and tails. In the language of the random walk, the more steps we take the less likely it is that we'll be at our starting point.

Also, after n coin flips, it is possible to have gotten n heads (or n tails). In terms of the random walk, this is equivalent to saying that it's possible to be n steps away from the starting point, in either direction. However, the standard deviation of the PDF only increases with the square root of n.

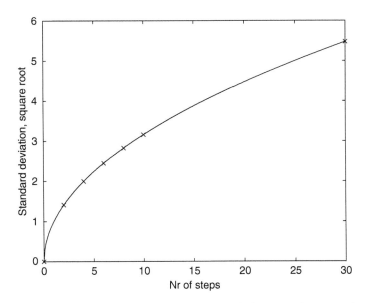

Figure 4.3 Standard deviation of distance traveled in a one-dimensional random walk.

The ratio of σ to n is

$$\frac{\sigma}{n} = \frac{\sqrt{n}}{n} = \frac{1}{\sqrt{n}} \tag{4.1}$$

The ratio σ/n gets smaller as n gets larger. For small n, it's hard to picture what a 95% confidence interval means. For large n, the random walk PDFs start looking normal and saying that the 95% confidence interval is $\pm 2\sigma$ about the mean is a good approximation.

Now, n is the farthest distance you could have walked. Therefore, if σ/n is getting smaller, than the region in which you are 95% confident of ending up after n steps is getting to be a smaller *fraction* of the furthest distance from your starting point that you might possibly have ended up.

Some Subsequent Calculations

The above argument holds for any confidence interval (number of σ). In other words, the region about your starting point in which you are any % confident of ending up after n steps is getting to be a smaller fraction of how far away you could possibly be as n increases.

σ is increasing as n increases. The 95% (or any other %) wide confidence factor region is getting larger as n increases. For a large value of n, you really have no idea where you'll be – you just know that it's highly unlikely that you'll be as far away as you possibly could be.

The above conclusions all pass common sense tests. If you flip a coin 4 times and get 3 heads, it's certainly not a big deal. If you flip a coin 400 times and get 300 heads, however, this doesn't look like an honest coin and the random walk based on these flips starts looking very directed. Similarly, if you flip a coin 20 times and get 18 more heads than tails (i.e. 19 heads and 1 tail), you should be suspicious. However, if you flip a coin 200 times and get 18 more heads than tails (109 heads and 91 tails), it's not exceptional. The exact probabilities of the above events can be calculated using the binomial probability formula, the subject of an upcoming chapter.

Now, we have seen that when we flip a coin n times (n even), the highest probability, or most likely event, is that we'll get $n/2$ heads and $n/2$ tails. As we increase n, i.e. flip the coin more and more times, even though it's less likely in an absolute sense that we'll get $n/2$ heads and $n/2$ tails, the ratio of the standard deviation to n falls. In other words, the confidence interval, for any measure of confidence, divided by n, falls.

Relating this to the random walk, the confidence interval about the expected location is telling us where (in what range of x) on the ruler we have a 95% probability of being. As n gets larger (more steps), both the standard deviation of x and the confidence interval grow. However, the confidence interval *as a fraction of the standard deviation of x* falls.

If we were to require that the 95% confidence interval be no more than, say, 10% of the number of coin flips, using Eq. (3.1) we can write

$$2\sigma = 2\sqrt{n} \leq .1n \tag{4.2}$$

Carrying through the algebra, we get $n \geq 400$. If we wanted the 95% confidence interval to be no more than 1% of the number of coin flips, then using the same calculation, we get $n \geq 40\,000$.

Summarizing, if you want to know something to a given confidence interval as a very small fraction of the number of data points you've got, you need a very large number of data points.

Table 4.2 shows an interesting way to look at (or calculate) the probabilities of this random walk. At the start, our walker is at $x = 0$. He hasn't gone anywhere yet, so his probability of being at $x = 0$ is 1.00. He takes his first step, either to the left or the right, with equal probabilities. This means that the probability of his being at $x = 1$ or $x = -1$ must be half of his previous probability, while his probability of being where he was ($x = 0$) is exactly 0. This is shown in the line *step 1*.

Getting to step 2 is slightly, but not very much, more complicated. From each of the positions he might have been at after step 1, he has half of his previous probability of moving 1 step to the right or 1 step to the left. Therefore, his probability of being at $x = 2$ or $x = -2$ is half of .500, or .250. Since he has 2 chances of arriving back at position 0, each with probability .250, his

Table 4.2 A one-dimensional walk, propagation of probabilities.

Nr steps					Probability				
0					1				
1				.500	0	.500			
2			.250	0	.500	0	.250		
3		.125	0	.375	0	.375	0	.375	
4	.0625	0	.250	0	.375	0	.250	0	.250
Position	−4	−3	−2	−1	0	1	2	3	4

probability of getting to position 0 is .250 + .250 = .500. Again, his probabilities of being where he had been at step 1 are exactly 0.

This process can be carried on forever. In each case, the probability at a given spot splits in half and shows up down one line and both one step to the left and one step to the right. When a point moves down and, say, to the right, and lands in the same place as another point moving down and to the left, then the two probabilities add.

Diffusion

The above process can be continued forever. When viewed this way, we may think of the probability (of being at a given point) as a propagating phenomenon, *smearing out* across space with advancing time. This perspective ties the idea of a random walk to the physical process called *diffusion*.

To illustrate a more typical perspective on diffusion while showing that it is indeed a random walk process, we'll construct an example of the propagation of probabilities using ants.

At time = 0, deposit a very large number of ants in a starting region. Put 1 million ants at $x = 0$, 1 million ants at $x = -1$, and 1 million ants at $x = +1$. Assume that each ant either stays still or wanders either 1 unit in the $+x$ or 1 unit in the $-x$ direction per second randomly, with an equal probability of one of these three choices. Since the number of ants is very large, we'll approximate the actual movements by their average value, i.e. one-third the ants move to the left, one-third the ants move to the right, and one-third of the ants remain in place.[2]

The ant distribution function (ADF) at after one time step in Figure 4.4a. At time = 0, there were 1 million ants at $x = -1$, $x = 0$, and $x = 1$ because we put

---

2 In the field of statistical mechanics, the movement of molecules in an *ideal gas* is approximated in just this manner.

them there. Between time = 0 and time = 1, there was an average motion of one-third million ants in both the +x and −x directions from each of these three locations. This means that there was no net flux of ants in to or out of $x = 0$. The only locations that see a net flux in or out are those locations where there is a difference in the starting numbers between those locations and adjacent locations.

Figure 4.4b and c shows the ADF after two time steps and three time steps, respectively.

There is only a net flow in any direction at a given point due to diffusion when there is a difference in the concentration of particles (i.e. a *concentration gradient*) on either side of the point. The net flow will always be in the direction from the higher concentration to the lower concentration. Diffusion occurs in gases and in liquids (and at very high temperatures in solids).

Continuing with our example, Figure 4.4d shows the ADF after 17 time steps.

The particles are spreading out, moving – or diffusing – away from their starting cluster and evolving their original uniform density to the familiar normal distribution. This is what happens (in three dimensions) when a few drops of cream are dripped into a cup of coffee, when some colored gas is introduced into a colorless gas, etc.

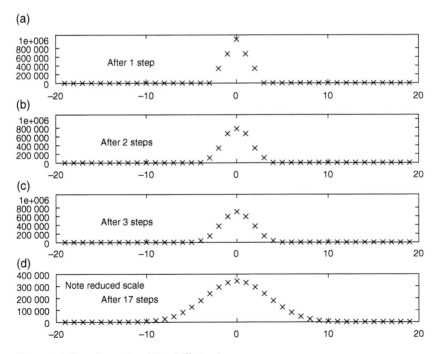

Figure 4.4 One-dimensional "Ant Diffusion."

If we continued this process indefinitely, we would see the normal distribution getter wider and wider. Since the total number of ants (aka the area under the curve) is a constant, the peak of the curve, at $x = 0$, would get lower and lower.

Now let's modify the situation by changing points $x = -18$ and $x = 18$ to reflecting walls. These walls have the property of bouncing back every ant (particle) that bumps into them. For <18 time steps this has no effect whatsoever, because no ants have reached these walls yet. This is the situation just after time step 17.

Figure 4.5 shows the ADF at $t = 25$, which is hardly different from the situation at $t = 19$, and then at several later times. The discrete points of the previous graphs have been replaced with continuous lines for clarity.

The distribution of ants in the box, or gas molecules in a box, or cream in a cup of coffee, or ..., diffuses from a high concentration over a small region to a lower uniform concentration over the entire box. The total number of ants, or ..., of course remains constant. In Chapter 17 it will be shown how the concepts of gas temperature, gas pressure, and the ideal gas are derived from this simple calculation (extended of course into three dimensions).

In the above example, we started with 3 million ants and have not added or lost any ants. At each time step, we approximated random motion by having 1/3 of the ants (1 million ants) go to the left one step, 1/3 of the ants go to the right one step, and 1/3 of the ants stay where they were. This means that, at each time step, 2 million ants are moving one step.

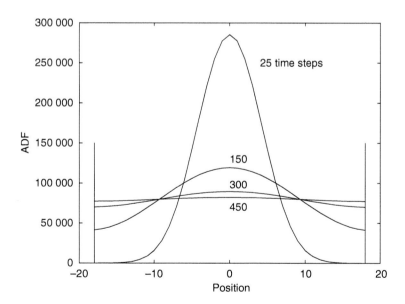

Figure 4.5 "Ant Distribution" functions in a box with reflecting walls.

Ultimately, the distribution of ants between the walls is uniform and does not change from time step to time step. At each time step, 2 million ants are moving. Because there are no concentration gradients, however, the net motion at any location is zero. This is exactly what happens in a closed room, for example. There are no breezes, the pressure (at a macroscopic level) is constant across the room while all the air molecules are in constant motion. If you were to hold up a small empty picture frame at any place in the room, there would constantly be air molecules passing through it, but the net transfer of molecules (the "flux") would be zero.

Returning to the properties of a random walk, in a random walk the steps are independent of each other. Each step is determined by (the equivalent of) the flip of a coin and each step has no knowledge whatsoever of what the previous steps were. The standard deviation of the location probability, σ, grows with the square root of the number of steps, $\sqrt{n}$.

This square root relationship is a general property of independent (random) events. For example, suppose we have an old clock with somewhat warped parts. The clock, on the average, keeps pretty good time, but at any given time its hands will predict a time that's distributed normally about the actual time, with a standard deviation of, say, three minutes. A similar clock, with even worse warpage, will also predict a time that's distributed normally about the actual time, with a standard deviation of four minutes. If we compare the clocks every 5 minutes for a full day, what will the distribution of the differences of the predicted times of these clocks be?

Since both clocks, on the average, tell the correct time, the average value of the difference between them will be zero. However, the standard deviation of these errors will be

$$\sigma_{difference} = \sqrt{3^2 + 4^2} = \sqrt{9 + 16} = \sqrt{25} = 5 \tag{4.3}$$

This is a smaller answer than you get by just adding $3 + 4 = 7$, because both clocks' errors are random, and there is some propensity (of these errors) to cancel. This is directly analogous to the random walk because since forward and backward steps are equally likely, they'll tend to cancel (that's why the expected value of distance traveled is 0). The standard deviation keeps growing with the number of steps, it grows as the square root of the number of steps.

If the steps, or clocks, or whatever, are not independent then *all bets are off*. For example, if our random walker is more likely to take a step in the same direction as his previous step because he doesn't like to turn about, then the steps are not independent. In this case it's impossible to predict the standard deviation without knowing the exact details of the dependence (i.e. the nonindependence) of a step on the previous step (or steps).

Problems

4.1 Consider a simple one-dimensional random walk. A walker will randomly move forward one step or back one step at each time interval. The path is on a slight hill so the probability of moving to the right is .6. Create the PDFs of the walker's position after each time step for four steps. (This may be done by hand, but it's easier to do on a spreadsheet.) Calculate the EV and σ after each step.

4.2 The PDF for Problem 4.1 does not look physical due to the alternating zeros that the algorithm creates. This was addressed in the ant diffusion example by smoothing the propagation – 1/3 of the ants moved in each direction, 1/3 stayed still at each time step. Repeat Problem 4.1 using this same strategy for probabilities, i.e. probability = .25 to move left, .3 to stay still, and .45 to move right.

4.3 A two-dimensional random walk is a random walk in which the walker has four choices of direction. Assume a walker is starting at $x = 0$, $y = 0$ and takes a step every second with the probability .2 of moving either in the $+x$, $-x$, $+y$, $-y$ directions or not taking a step. Calculate the location probabilities for the first four steps. If you have a spreadsheet or computer software with three-dimensional surface plotting capability, plot the location probabilities after four steps.

4.4 Figure P4.4 depicts the geometry for an ant-in-a bowl random walk and it shows a parabola, $-1 \le x \le 1$, with an ant sitting at the right edge of it.

The ant moves in a simple .1-length-step per second random walk, but the probability of the direction changes with the ant's position on the parabola: If it's near the center, the chosen direction is random, as it moves up one of the slopes, it develops a propensity to move back down the slope. The probability of a step (for motion to the left or to the right) is given by

$$P_{left} = .5 + .5x \qquad x \ge 0$$
$$P_{right} = 1 - P_{left}$$

$$P_{right} = .5 - .5x \qquad x \le 0$$
$$P_{left} = 1 - P_{right}$$

A Start the ant at $x = 1$ and track its motion for a while.
B Discuss the probability of the ant returning to $X = 1$

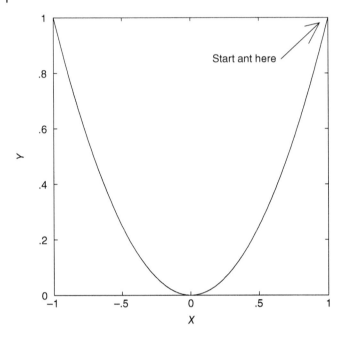

Figure P4.4 Bowl geometry for ant random walk calculation.

Notes:
1 Since a random number generator is involved, no two solutions will be exactly the same.
2 We're not following Newton's laws, as a marble rolling in the parabola would.
3 This problem is awkward by hand, OK on a spreadsheet, but best done using a programming language of your choice.

4.5 Repeat Problem 4.4, but do it as a diffusion problem, with 21 locations spaced .1 apart, $-1 \leq x \leq 1$. Start 1 000 000 ants evenly divided between $x = 1$ and $x = .9$, use the existing probability numbers as "fraction of group that moves each way" numbers. If you're working by hand, just take as many steps as you have patience for. If you're working with a spreadsheet or with a programming language, take enough steps to see if the ant population distribution stabilizes or oscillates indefinitely.

4.6 Figure P4.6 shows the geometry of a "room" for a Monte Carlo ant walk.
The figure shows a rectangular box with two smaller rectangular boxes inside it. The outer box has dimensions 1000×1000. The inner rectangles have locations and dimensions (locations referenced to the lower left corner of the outer box)

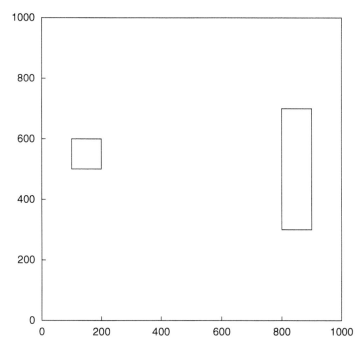

Figure P4.6 Floor with two holes geometry for random walk Monte Carlo calculation.

Left Box: lower left = (100,500), upper right = (200,600)
Right Box: lower left = (800,300), upper right = (900,700).

We drop an ant in the center of the outer box (500,500). The ant wanders in a simple two-dimensional random walk. If it encounters the outer wall it reverses direction; if it encounters one of the inner boxes it "falls off the edge of the earth" and its walk is over.
A Trace at least one path for an ant.
B Launch enough ants to predict the probability of an ant ending up at either of the inner boxes.

5

Life Insurance

Introduction

In this chapter we'll introduce the concept of insurance, using life insurance as our example. We'll also mention US Social Security. It is important to emphasize that the dollar numbers for Social Security change often and that the numbers used in the examples in this chapter are only approximations.

In order to calculate the cost of an insurance policy we need to introduce the financial concept of Present Value. While insurance premiums are definitely a probability-based calculation, present value is not probability based or related. However, it is needed to bring meaning to insurance premium calculations.

Life Insurance

Insurance in its various forms is something that all of us deal with at some time or other. There are so many different kinds of insurance policies that it's impossible to consider discussing them all. We will concentrate on simple life insurance policies. Most people don't actually buy simple life insurance policies; insurance companies like to put together combinations of life insurance policies, savings accounts, annuities, and investment portfolios into big packages. There's nothing inherently wrong with doing this provided you understand what it is you're buying, what the terms of each of the components of the package are, and how all of these components interact.

There are also different forms of simple life insurance policies as well as different names for the same life insurance policy. The simplest life insurance policy, which will be discussed in detail below, is the *term* life insurance policy. When you buy a term policy, you pay a certain amount up front (the *premium*). If you die during the specified term of the policy, the insurance company pays

Probably Not: Future Prediction Using Probability and Statistical Inference,
Second Edition. Lawrence N. Dworsky.
© 2019 John Wiley & Sons, Inc. Published 2019 by John Wiley & Sons, Inc.
Companion website: www.wiley.com/go/probablynot2e

a fixed amount that is specified in the policy. If you survive the term of the policy, you and the insurance company no longer have a relationship; they keep the premium, they owe you nothing ever, unless of course you purchase another policy. A simple variation of the term policy is the *decreasing* term policy. You pay a premium up front and if you die during the term of the policy the insurance company pays off. However, the amount of the payoff decreases in steps over the term of the policy. A common decreasing term policy is the mortgage insurance policy. The payoff amount of the mortgage insurance policy is tied to the balance of the mortgage on your home, which decreases as you make your monthly mortgage payments. Also, you typically pay the premium periodically rather than up front.

Real insurance companies, like all other real companies, are in business to make money. Along the way, they have a cost of doing business which comes not just from paying out on policies but also from employee salaries, taxes, office equipment, their office real estate, etc. The simple rule is that they must take in more than they pay out to policy holders in order for them to stay in business. Business finance is a topic well worth studying (especially if you're running or planning to run a business). However, this book is about probability and this chapter is about how probability calculations determine life insurance policy costs. We'll therefore leave the business issues to books about business issues; our insurance companies are idealized operations staffed by volunteers. They have no costs other than policy payouts, they make no profits, and they take no losses. The expected value of their annual cash flow is zero. This idealization lets us deal with the probability concepts inherent in insurance policies without getting entangled in the business issues of running a real company.

Insurance as Gambling

Suppose that you walk into your bookie's "office" and ask "What is the probability that I will die of a cause other than suicide in the next 12 months?" The bookie knows you to be fairly young, healthy, he's sure that you have no diseases that you're hiding. He stares at the ceiling for a while and/or consults his crystal ball, and replies: "one percent." You and your bookie then make a bet: you give him $1000. He gives you his *marker*, which is a note that says something like "If you die for any reason (other than suicide) within 12 months of today, I will give $100,000 to you – actually to your family, considering the nature of the bet. In either case, I keep the $1,000."

Most of us really do handle things this way; we just change the names of the items and the title of our bookie. The more common scenario is to walk into your insurance broker's office and ask: "What is the probability that I will die of a cause other than suicide in the next 12 months?" Your insurance broker asks you to fill out some forms, possibly to get a physical examination. She then stares at her Actuarial, or Life, Tables for a while and replies "one percent." Once again, we'll assume that she is willing to take the bet; that is, to write you a policy. You give her $1000. She gives you *her marker*, which in this case is labeled "Term Life Insurance Policy." The insurance policy, boiled down to its simplest terms, says something like "If you die for any reason other than sui-cide within 12 months of today, the insurance company will give $100000 to you – actually, to people called beneficiaries – considering the nature of the policy." Whether you die during this period or not, the insurance company keeps the $1000.

When looked at this way, it becomes clear that both gambling casinos and insurance companies are in the same business: taking long odds bets from people who come through the door. Gambling casinos usually deal with slot machine kinds of bets while insurance companies usually deal with insurance kinds of bets, but they both have very sophisticated sets of tables telling them the probabilities on every bet that they might take. In the case of the gambling casino, these tables take the form of probabilities based on the mechanisms in the slot machines and the number layout of the roulette wheels; in the case of the insurance company these tables take the form of average life expectancies, etc. Since terms such as *betting* and *gambling* tend to make people nervous, whereas *having insurance* tends to make people feel warm and secure, insur-ance companies have their own language and are careful to minimize any refer-ences to betting and odds. From the points of view of the gambling casino and the insurance company, the difference between the two different types of bets is totally cosmetic.

Let's examine the purchase of a life insurance policy. Assuming that you're in reasonable mental and physical health and are living a fairly happy life, you really don't want to win this bet. That is, you don't want to die sometime in the next 12 months. Fortunately, in the above example, the odds are 99 to 1 against you dying in the next 12 months.

We must now confront a very perplexing situation: you have placed a bet where the odds are 99 to 1 against you winning, and furthermore, *you really don't want to win this bet.* (Remember that to win this bet you must die some-time in the next 12 months.) From this perspective, placing the bet doesn't seem like a good idea at all. Why then did you place this bet? Why do so many of us place this bet; that is, buy life insurance policies?

The answer lies in a personal application of a philosophy called the *Precautionary Principle*: a reasonable person will bet against very long odds, on a bet that he really does not want to win, if the side effects of win-ning would be absolutely unacceptable without the bet component. In the case of life insurance, absolutely unacceptable could mean that if the wage earner of a family dies, the family would not have the financial resources to carry on. Having a life insurance policy to protect against this situation is a very good idea.

The Precautionary Principle as you would find it in many sites on the internet usually deals with scientific and/or political decisions affecting public issues such as the environment and public safety. There are many definitions cited and even discussions comparing and contrasting these different definitions.

An example of these definitions is "… a willingness to take action in advance of scientific proof [or] evidence of the need for the proposed action on the grounds that further delay will prove ultimately most costly to society and nature, and, in the longer term, selfish and unfair to future generations."[1] It's not hard to imagine how this societal definition evolved from a personal definition such as the one given above and how the personal definition in turn evolved from our evolved pattern-generating "there might be tigers in the bush" response discussed in Chapter 1.

Returning to the difference between gambling and buying insurance, our first conclusion is that from the perspective of the gambling casino owner and the insurance company, there is absolutely no difference. In both cases, there are a lot of customers; so it's unlikely that there will be too much deviation from the actuarial tables/odds. Each costumer is paying a small amount as compared to the payback: another way of saying that these are "long odds." In practice – of course – the gambling casino and the insurance company skew the price : payoff ratio somewhat so as to assure a steady flow of money into their pockets to cover costs of running the system and profits. In both cases, there are many attractive products on sale; so it's pretty certain that each customer will in reality pay more than a "small amount." In the case of the gambling casino, there are slot machines, roulette wheels, craps tables, etc. In the case of the insurance company, there is health insurance, auto insurance, etc., and of course the many different life insurance-based products.

Should we conclude that there's absolutely no difference between placing a gambling casino bet and buying an insurance policy? No. There is one very big difference which is implicit in the discussion above. When you put money into a slot machine you're hoping to win the bet. When you pay a year's premium on a life insurance policy you're hoping to lose the bet but cannot accept the consequences of winning the bet unless you have placed a big enough bet. In other words, you're hoping you don't die during the next 12 months, but just in case you do die, you want to make sure you have enough life insurance, so that your heirs are left economically secure. This simple but very important difference makes gambling a frivolous and sometimes painful entertainment while prudent insurance purchases are an important part of personal and family financial security and health planning.

---

1 O'Riordan, Tim and Cameron, James (1994). *Interpreting the Precautionary Principle.* Earthscan Publications Ltd. ISBN: 1-85383-200-6.

Life Tables

In order to calculate the average costs of writing life insurance policies, we need to know the probabilities of people dying at different ages. This information is presented in Life Tables. Life Tables are available online, along with tutorials on how they're calculated, precise definitions of the terms used, etc. Tables are calculated and published for various subgroups of the population such as sex and race as well as for the entire population. These tables must be updated frequently because advances in medical care, nutrition, demographics, etc., cause changes to the data.

Table 5.1 shows the data for the entire population of the United States in the year 2014. The leftmost column, x, is the age at (your) last birthday.[2] The data is presented in one-year increments. At the bottom of the table you can see that there is no specific data for people more than 113 years old, their data is simply grouped into a 113+ catch-all category.

The second column, q_x, is the probability of dying during year x. The year of birth ($x = 0$, the top row in the table) is a relatively dangerous year. The probability of dying actually falls with age for a few years after birth, then starts rising again, but doesn't exceed the birth year probability until the 55th year. Being born is a risky process: there are issues of birth defects leading to an unsustainable infant life, problems occurring during birth, etc.

Note also that q_{113}, the catch-all probability that (if you reach age 113) you will certainly die at some age greater than 113, is 1.000, the certain event. The reasoning behind this should be obvious.

The numbers in the second column are independent of each other, they are not cumulative. That is, the probability of you dying in your 20th year assumes that you have succeeded in reaching your 20th year.

The third column, "Number Surviving to Age X," starts with 100 000 people. This number has nothing to do with the population of the United States in 2014. It is just a convenient starting number from which we can calculate fractions of the population, i.e. if we have 99 313 people this year and had 100 000 people last year, then 99 313/100 000 = 99.313% of the people who were alive last year are still alive (entering) this year.

The third column (labeled l_x) is the number of people (survivors) reaching the starting age for a given row (x) assuming that we began with 100 000 people. In order to celebrate your 10th birthday, you must have already celebrated your first, second, third, …, and ninth birthdays. These numbers are therefore cumulative and since the result at each year depends on the result for the previous year, the relationship is referred to as recursive. Since $1 - q(x)$ is the probability of Not dying in year x,

$$l_{x+1} = l_x \left(1 - q_x\right) \tag{5.1}$$

---

2 This age is the answer you would give to the question "How many years old are you?"

Table 5.1 Life tables.

Age	Death prob	Nr surviving to age x	Nr dying at age x	Nr person-years	Person-years above age x	Expectation at age x
x	q_x	l_x	d_x	L_x	T_x	e_x
0	.005815	100 000	582	99 709	7 872 204	79
1	.000377	99 419	38	99 400	7 772 495	78
2	.000252	99 381	25	99 369	7 673 095	77
3	.000186	99 356	19	99 347	7 573 727	76
4	.000156	99 338	16	99 330	7 474 380	75
5	.000136	99 322	14	99 315	7 375 050	74
6	.000126	99 309	13	99 302	7 275 735	73
7	.000116	99 296	12	99 290	7 176 433	72
8	.000106	99 285	11	99 279	7 077 143	71
9	.000086	99 274	9	99 270	6 977 863	70
10	.000096	99 266	10	99 261	6 878 594	69
11	.000091	99 256	9	99 252	6 779 333	68
12	.000116	99 247	12	99 241	6 680 081	67
13	.000166	99 236	17	99 227	6 580 840	66
14	.000227	99 219	23	99 208	6 481 613	65
15	.000297	99 197	30	99 182	6 382 405	64
16	.000368	99 167	37	99 149	6 283 223	63
17	.000444	99 131	44	99 109	6 184 075	62
18	.000525	99 087	52	99 061	6 084 966	61
19	.000611	99 035	61	99 004	5 985 906	60
20	.000697	98 974	69	98 940	5 886 901	59
21	.000773	98 905	77	98 867	5 787 962	59
22	.000850	98 829	84	98 787	5 689 095	58
23	.000891	98 745	88	98 701	5 590 309	57
24	.000912	98 657	90	98 612	5 491 608	56
25	.000933	98 567	92	98 521	5 392 997	55
26	.000965	98 475	95	98 427	5 294 476	54
27	.000981	98 380	97	98 331	5 196 049	53
28	.001012	98 283	100	98 233	5 097 718	52
29	.001049	98 184	103	98 132	4 999 485	51

Table 5.1 (Continued)

Age	Death prob	Nr surviving to age x	Nr dying at age x	Nr person-years	Person-years above age x	Expectation at age x
x	q_x	l_x	d_x	L_x	T_x	e_x
30	.001086	98 081	107	98 027	4 901 353	50
31	.001118	97 974	110	97 919	4 803 325	49
32	.001160	97 865	114	97 808	4 705 406	48
33	.001212	97 751	119	97 692	4 607 598	47
34	.001260	97 633	123	97 571	4 509 907	46
35	.001318	97 510	129	97 445	4 412 336	45
36	.001391	97 381	136	97 313	4 314 890	44
37	.001465	97 246	143	97 174	4 217 577	43
38	.001545	97 103	150	97 028	4 120 403	42
39	.001624	96 953	158	96 874	4 023 375	41
40	.001725	96 796	167	96 713	3 926 501	41
41	.001847	96 629	179	96 539	3 829 788	40
42	.001980	96 450	191	96 355	3 733 248	39
43	.002150	96 259	207	96 156	3 636 894	38
44	.002342	96 052	225	95 940	3 540 738	37
45	.002551	95 827	245	95 705	3 444 799	36
46	.002788	95 583	267	95 449	3 349 094	35
47	.003063	95 316	292	95 170	3 253 645	34
48	.003368	95 024	320	94 864	3 158 475	33
49	.003701	94 704	351	94 529	3 063 611	32
50	.004080	94 354	385	94 161	2 969 082	31
51	.004464	93 969	420	93 759	2 874 921	31
52	.004874	93 549	456	93 321	2 781 162	30
53	.005312	93 093	495	92 846	2 687 841	29
54	.005767	92 599	534	92 332	2 594 996	28
55	.006251	92 065	576	91 777	2 502 664	27
56	.006777	91 489	620	91 179	2 410 887	26
57	.007313	90 869	665	90 537	2 319 708	26
58	.007821	90 205	706	89 852	2 229 172	25

(*Continued*)

Table 5.1 (Continued)

Age	Death prob	Nr surviving to age x	Nr dying at age x	Nr person-years	Person-years above age x	Expectation at age x
x	q_x	l_x	d_x	L_x	T_x	e_x
59	.008363	89 499	749	89 125	2 139 320	24
60	.008946	88 751	794	88 354	2 050 195	23
61	.009613	87 957	846	87 534	1 961 842	22
62	.010297	87 111	897	86 663	1 874 308	22
63	.011025	86 214	951	85 739	1 787 645	21
64	.011805	85 264	1007	84 760	1 701 907	20
65	.012705	84 257	1071	83 722	1 617 146	19
66	.013716	83 187	1141	82 616	1 533 425	18
67	.014864	82 046	1220	81 436	1 450 809	18
68	.016146	80 826	1305	80 174	1 369 373	17
69	.017587	79 521	1399	78 822	1 289 199	16
70	.019265	78 123	1505	77 370	1 210 378	15
71	.021144	76 618	1620	75 808	1 133 008	15
72	.023161	74 998	1737	74 129	1 057 200	14
73	.025307	73 261	1854	72 334	983 071	13
74	.027624	71 407	1973	70 420	910 738	13
75	.030324	69 434	2106	68 381	840 317	12
76	.033433	67 329	2251	66 203	771 936	11
77	.036902	65 078	2402	63 877	705 733	11
78	.040733	62 676	2553	61 400	641 856	10
79	.044999	60 123	2706	58 770	580 457	10
80	.049950	57 418	2868	55 984	521 687	9
81	.055583	54 550	3032	53 034	465 703	9
82	.061736	51 518	3181	49 927	412 670	8
83	.068405	48 337	3307	46 684	362 742	8
84	.075760	45 031	3412	43 325	316 059	7
85	.083988	41 619	3496	39 871	272 734	7
86	.093289	38 124	3557	36 345	232 863	6
87	.103784	34 567	3588	32 773	196 517	6
88	.115528	30 980	3579	29 190	163 744	5

Table 5.1 (Continued)

Age	Death prob	Nr surviving to age x	Nr dying at age x	Nr person-years	Person-years above age x	Expectation at age x
x	q_x	l_x	d_x	L_x	T_x	e_x
89	.128538	27 401	3522	25 640	134 554	5
90	.142764	23 879	3409	22 174	108 915	5
91	.158162	20 470	3238	18 851	86 741	4
92	.174733	17 232	3011	15 727	67 890	4
93	.192286	14 221	2735	12 854	52 163	4
94	.210900	11 487	2423	10 275	39 310	3
95	.229479	9064	2080	8024	29 034	3
96	.247709	6984	1730	6119	21 010	3
97	.265417	5254	1395	4557	14 891	3
98	.281902	3860	1088	3316	10 335	3
99	.297492	2772	825	2359	7019	3
100	.313559	1947	611	1642	4660	2
101	.331089	1337	443	1115	3018	2
102	.349553	894	313	738	1903	2
103	.368874	582	215	474	1165	2
104	.389646	367	143	296	691	2
105	.412946	224	93	178	395	2
106	.433460	132	57	103	218	2
107	.463087	75	35	57	115	2
108	.487500	40	20	30	57	1
109	.512195	21	11	15	27	1
110	.550000	10	6	7	12	1
111	.555556	5	3	3	5	1
112	.750000	2	2	1	1	1
113+	1.000000	1	1	0	0	0

l_1 (age = 0) is easy to understand. It says that everyone who is born must be alive on the day that they're born. Since we're starting with a population of 100 000, l_1 must be 100 000. Again, remember that 100 000 is just a convenient number to start with so that we can track deaths and remaining population over the years. Now, as pointed out above, if q_1 is the probability of someone dying during their first year of life, then $1 - q_1$ must be the probability of someone NOT dying during this period, i.e. making it to their first birthday. If we are starting with 100 000 people and the probability of each of them reaching their first birthday is

$$1 - q_1 = 1 - .005815 = .99419 \tag{5.2}$$

then

$$l_1(1 - q_1) = 100000(.99419) = 99\,381 \tag{5.3}$$

of these people will be alive on their first birthday, and this is the value of l_2.

The rest of the l column is built following this procedure, each result being used to calculate the next.

The fraction of people alive at the start of a given year of life (x) is just $l_x/100\,000$. For example, the number of people reaching their 25th year is just .985657 or approximately 98.6% of the people born 25 years earlier. Now between their 25th and 26th years, people will die at a more-or-less uniform rate. However, just to keep things simple let's assume that 98.6% of the original group of people is the average number of people alive during their 25th year. The conclusions reached in the calculations below will therefore not be perfectly accurate, but the points to be made will be clearer.

Birth Rates and Population Stability

Before going through the rest of the columns in the Life Table, let's see what we can learn from what we have so far. In order for there to be 100 000 25-year-old people alive, the starting group must have had 100000/.9857 = 101 450 babies born, or 1.01450 more people born than reaching their 25th birthday. Assuming that half of the population is women, and that they all arranged to give birth during, and only during, their 25th year, each woman would have had to give birth to 2(1.01457) = 2.029 babies. This is of course a somewhat nonsensical perspective. Women who have children do not have them all in their 25th year, and they certainly do not average twins or more per birth. Births occur, statistically, over a range of (women's) lives. The point here is that it is not enough for each woman, on the average, to give birth to a total of 2 children, or even for each couple, on the average, to give birth to 1 child per person, for the population to be stable. The l_x column of the Life Table decreases with each year. Other than by immigration of people into the United States, there is no way to replace people who die each year, and in order for the population to be stable the average number of births per woman must be greater than two.

Returning to the simple assumption of all births taking place in one year, what if we change that year from age 25 to age 40? Looking at the Life Table and repeating the calculation, we find that the average number of births per woman jumps from about 2.029 to about 2.067. This makes sense because some women have died between ages 25 and 39, and therefore a higher average birth rate per mother is needed to maintain the population. This effect has actually been happening in the United States because more and more women want to have both families and established careers and are consequently starting families later in life than the earlier generations. Advances in medical care that make this decision safe have also certainly influenced the decision.

The actual birth rate is based upon a weighted average of all the birth rate statistics. However, the bottom line is still that an average birth rate of greater than 2 babies per mother, over the mother's life, is necessary to maintain a constant total population. If the birth rate is lower than this number, the population will continue to fall; if the birth rate is higher than this number, the population will continue to grow. In the former case, the population of people will eventually disappear. In the latter case, eventually the number of people will outpace the ability to feed these people and the Life Tables (due to changes in the q column) will change. Over the course of history there have been many factors influencing the Life Tables in addition to the simple availability of food: wars, disease, religious beliefs and practices, etc., all weigh in. However, as will be shown below, the ability to accurately predict the cost of a life insurance policy many years in advance depends on the stability of the Life Tables, or at least on a good picture of how they're going to change over the years. This of course is just a restatement of everything we've looked at so far – if the coin you're about to flip is continually being replaced with another coin with a new (and unknown) probability of landing on heads, you can't know what your betting odds should be to keep the game fair.

Life Tables, Again

Returning to the Life Table, note that all of the information in the table is based on the first column (age, or x) and on the second column, q_x. Everything else is calculated from these first two columns.

The fourth column, d_x, is the number of people dying at a given age. That is,

$$d_x = l_{x+1} - l_x \tag{5.4}$$

This is a fairly straightforward calculation to understand. If 100 000 people are born and only 99 313 reach their first birthday, then 100 000 − 99 313 = 687 people must have died between birth and their first birthday.

The fifth column, L_x, is the number of person-years lived between ages x and $x + 1$. This is a calculation we'd need if, for example, we knew that the "average person" of age x needs 2 quarts of water a day and wanted to know how much water we would need to supply a large group for a year.

If we assume that people die at a uniform rate, then we would expect we get L_x by averaging l_x and l_{x+1}, that is,

$$L_x = \frac{l_x + l_{x+1}}{2} \tag{5.5}$$

Equation (5.5), which was used to generate L_x in Table 5.1, implicitly assumes that deaths are uniformly distributed across the year. This is reasonable for all years except $x = 0$, the year of birth. In this case the deaths are, statistically, *bunched up* at the beginning of the year (at birth) and the table entry is a poor approximation to the actual number.

The sixth column, T_x, is the total number of person-years lived above age x. These numbers are calculated by starting at the bottom of the table and working up, that is

$$T_x = T_{x+1} + L_x$$

Starting at the bottom of the table, T_{113} must equal L_{113} because $x = 113$ is our final "catch-all" row in the table. What this assertion is saying is that the total number of people-years lived between age 113 and all higher ages (L_{113}) must be the same as the total number of people-years lived above age 113 (T_{113}). Then, the total number of people-years lived above age 112 (T_{112}) must just be the sum of the total number of people-years lived *above* age 113 (T_{113}) and the total number of people-years lived between age 112 and age 113 (L_{112}). Follow this recursive procedure all the way up to the top of the table and you'll have all of the T values.

T_x is useful for predicting the amount of resources our population will need for the rest of their lives. If, for example, we want to give every 65-year-old person a certain pension and continue it until they die, how much money should we anticipate needing, on the average, per person? At age 65, we have 84 257 people (l_{65}). We are predicting 1617.146 person-years to come (T_{65}), so we should anticipate having to pay these people for an average of $T_{65}/l_{65} = 19.2$ years.

This last calculation is shown in the seventh and last column, e_x = expectation of life at age x:

$$e_x = \frac{T_x}{l_x} \tag{5.6}$$

When you're at some age x, e_x tells you how many more years you can expect to live. Your expected age at death is then just $x + e_x$.

Looking at the e_x column, we see that at birth the average life expectancy is 79 years, whereas at age 65 it is $65 + 19 = 84$ years. This seems very different than the situation we have when flipping a coin. The coin has no memory. If you just had the greatest run of luck ever recorded, that is, you just flipped a coin 100 times and got 100 heads, the expected result of the next 10 flips is still 5 heads and 5 tails. Why should your life expectancy behave differently – why should it change as you age?

The answer to this seeming paradox is that *your* personal life expectancy hasn't changed at all. What has happened is that the population base for the statistics has changed – some babies have succumbed to birth defects and early childhood diseases, some teenagers have gotten drunk and driven into trees, etc. By the time you reach 65 years of age, all the people who, for whatever reason, were to die at a younger age have already died and those that are left form a group with a longer life expectancy. This is also called *biased sampling*, and is logically not that different from the "marbles on a board" discussion in Chapter 1.

Let's look at a few other interesting numbers we can glean from these tables. The average age of the population is just the average of all ages weighted by the number of people alive at that age, l_x. Again, we realize that we are introducing a small error by assuming that l_x is the average number of people alive during the year x and also by the lumping together of everyone 113 or more years old, but as a good approximation, we calculate

$$avg\ age = \frac{\sum_{x=0}^{113} x l_x}{\sum_{x=0}^{113} l_x} = 41\ years \tag{5.7}$$

The standard deviation of this x is 24.8 years. The standard deviation is more than half of the average. Not many of the people are about 41 years old. Also, note that two standard deviations is 49.6 years. This is a case where the distribution is clearly Not Normal, there are no people less than 0 years old. If we want a confidence interval about the average age, we have to tabulate the CDF, we cannot simply count the number of standard deviations.

To calculate the median age, we sum the l_x numbers from $x = 0$ down and from $x = 113$ up, and watch for the point where these sums cross. In other words, at what age do we have as many people younger than that age as we do people older than that age? Without showing the table entries, this turns out to be 39 years old. This is approximately the same as the average age. The mode age is defined as the age at which we have the largest population. This is of course $x = 0$ years, a number totally unlike the mean or the median. Again, non-normal distributions can give very non-normal calculation results.

Premiums

Now let's look into the calculation of life insurance premiums. Remember that we're ignoring costs, profit margins, etc., and just treating life insurance premium calculations as a fair game gambling exercise. If you're just at your 65th birthday and you want to buy a year's worth of term insurance, then by looking at the q_x value (column 2) for $x = 65$, you see that your premium should just be $.0127 for every $1.00 of insurance. This is because if your probability of dying during the year between your 65th and 66th birthdays is .0127, then this is exactly the amount that the insurance company must charge for every dollar they would pay if you "won" the bet in order to make the game fair. A $100000 1-year term life insurance policy should cost $1270. However, remember that in practice, in addition to administrative costs and profits, your life insurance company will also be looking at tables that are much more specific to your age, sex, current health, etc., than our table – so don't expect to actually see this number.

Let's improve the above calculation a bit. $l_x = .0127$ isn't really the probability of dying at any time in your 65th year. The actual value must start (right at your 65th birthday) a bit higher than this and end (right before your 66th birthday) a bit lower than this so that an actual l_x curve calculated day-by-day rather than a year at a time is a smooth curve. The (q_x) value does not take a sudden jump the day of your 66th birthday. Let's assume that this value is an average value for the year and by working with this average we'll get a reasonable – if not as accurate as possible a – result.

Using this same reasoning, we have to look at the cost of the term policy. When you write a check and hand it to your insurance broker, you are loaning the insurance company money that they don't have to repay until the day you die (or not at all if you don't die during the term of the policy). Assuming that you will die during your 65th year, the average time that the insurance company will be holding your money is ½ year. This is not the same as a coin-flip game where money changes hands every minute or so. The fair game calculation must now consider the *present value* of $100000 returned to you ½ year after you gave it to the insurance company.

Present Value is a simple but important financial calculation: using your best estimate of the interest that you could earn on your money, how much money would you have to put into a savings account today so that you could withdraw $1270 a half year from now? Assume that the bank compounds interest twice a year. This means that if 3% is the interest rate, the bank will give you 1.5% interest on your money twice a year. Your cost for the insurance should therefore be reduced by the same factor, that is,

$$Premium = \frac{1270}{1.015} = 1252 \tag{5.8}$$

What about the cost of buying a $100000 policy on your 65th birthday that will cover you for the rest of your life? One way to solve this problem is to calculate the present value (on your 65th birthday) of one-year term policies for each year from your 65th to your "113+" years, and then add up the cost of all of these policies. There is an important difference between this proposed policy and the one-year term policy discussed above. In the case of the term policy, if your probability of dying during the term of the policy is, say, .01, then there's a .99 probability that the insurance company will get to keep your money and never pay out. In the case of a policy that covers you from a start date until your death, there is no doubt that at some time the insurance company will have to pay. This latter condition is reflected in the q_x column of the life table for age 100+ as a probability of 1.00, the certain event. As the old saying reflects, "you can't get out of this alive."

Consider Table 5.2. The first 4 columns are copied directly out of the Life Table shown earlier, but only the rows for $x = 65$ and higher are shown. The 4th column (d_x) shows the number dying during each year, based upon the original 100 000 people born. In order to turn this into a PDF, we must *normalize* this column, i.e. add up all of the entries and then divide each entry by this sum. What the normalization procedure does is to scale all the numbers so that they add up to 1. This result is a PDF for the probability of your dying in at a given age assuming that you've reached your 65th birthday. The results of this calculation are shown in column 5. Note that the first entry in this column is the same as the entry in the 2nd column (l). This is because both of these entries represent the same number: the probability of you dying during your 65th year, assuming that you are alive on your 65th birthday. All of the other entries in this column differ from the entries in the second column because the second column is the probability of dying at age x assuming that you have reached your xth birthday whereas the fifth column is the probability of you dying at age x looking forward from your 65th birthday. For example, look at the bottom row, $x = 113+$: the second column has probability of 1. This means that it is very certain that once you have reached your 113th birthday, you are going to die sometime after your 113th birthday. The fifth column has a probability of ~6/million. This means that looking forward from your 65th birthday, there is approximately a 6/million probability that you will live past your 113th birthday.

Figure 5.1 shows this (5th) column. Remember that this is the probability of your dying at a given age, looking from the vantage point of your 65th birthday.

Continuing, column 6 shows the number of years from your 65th birthday to the middle of year x. The argument here is the same as was given above in the term insurance example; we are assuming that the (insurance company's) bank compounds interest every half year and that the deaths all occur at mid-year. The seventh column shows the correction factor on the $100 000 payoff based on the number of years that the insurance company gets to keep the money. This again is just an extension of the term insurance calculation:

$$\frac{cost}{\$100\,000} = (1+.015)^{2n} \tag{5.9}$$

where n is the number of ½ years from your 65th birthday to your death and we are again assuming 3% annual interest compounded twice a year.

Table 5.2 Term life insurance premiums starting at 65th birthday.

Age	Nr surv Death P	Nr die At age x	At age x	d_x	Pmt Adv	Prem	Cost at 65th
X	q_x	l_x	d_x	Norm	Yrs	Corr	B'day
65	.012705	84257	1071	.012705176	.5	1.0149	1252
66	.013716	83187	1141	.013541902	1.5	1.0226	1324
67	.014864	82046	1220	.014473575	2.5	1.0379	1394
68	.016146	80826	1305	.015488327	3.5	1.0535	1470
69	.017587	79521	1399	.016598027	4.5	1.0693	1552
70	.019265	78123	1505	.017862017	5.5	1.0853	1646
71	.021144	76618	1620	.019226889	6.5	1.1016	1745
72	.023161	74998	1737	.020615498	7.5	1.1181	1844
73	.025307	73261	1854	.022004106	8.5	1.1349	1939
74	.027624	71407	1973	.023410518	9.5	1.1519	2032
75	.030324	69434	2106	.024989022	10.5	1.1692	2137
76	.033433	67329	2251	.026715881	11.5	1.1868	2251
77	.036902	65078	2402	.028502083	12.5	1.2046	2366
78	.040733	62676	2553	.030300153	13.5	1.2226	2478
79	.044999	60123	2706	.032110092	14.5	1.2410	2588
80	.049950	57418	2868	.034038715	15.5	1.2596	2702
81	.055583	54550	3032	.035985141	16.5	1.2785	2815
82	.061736	51518	3181	.037747606	17.5	1.2976	2909
83	.068405	48337	3307	.03924303	18.5	1.3171	2979
84	.075760	45031	3412	.040489218	19.5	1.3369	3029
85	.083988	41619	3496	.041486167	20.5	1.3569	3057
86	.093289	38124	3557	.042210143	21.5	1.3773	3065
87	.103784	34567	3588	.042578065	22.5	1.3979	3046
88	.115528	30980	3579	.042477183	23.5	1.4189	2994
89	.128538	27401	3522	.041800681	24.5	1.4402	2902
90	.142764	23879	3409	.040459546	25.5	1.4618	2768
91	.158162	20470	3238	.038424107	26.5	1.4837	2590
92	.174733	17232	3011	.035735903	27.5	1.5060	2373
93	.192286	14221	2735	.032454277	28.5	1.5286	2123

Table 5.2 (Continued)

	Nr surv	Nr die					
Age	Death P	At age x	At age x	d_x		Pmt	Prem	Cost at
					Adv		65th
X	q_x	l_x	d_x	Norm	Yrs	Corr	B'day
94	.210900	11 487	2423	.02875132	29.5	1.5515	1853
95	.229479	9064	2080	.024686376	30.5	1.5748	1568
96	.247709	6984	1730	.020532419	31.5	1.5984	1285
97	.265417	5254	1395	.016550554	32.5	1.6224	1020
98	.281902	3860	1088	.012912874	33.5	1.6467	784
99	.297492	2772	825	.009785537	34.5	1.6714	585
100	.313559	1947	611	.007245689	35.5	1.6965	427
101	.331089	1337	443	.005251789	36.5	1.7219	305
102	.349553	894	313	.003708891	37.5	1.7477	212
103	.368874	582	215	.002545783	38.5	1.7740	144
104	.389646	367	143	.001697188	39.5	1.8006	94
105	.412946	224	93	.001097832	40.5	1.8276	60
106	.433460	132	57	.000676502	41.5	1.8550	36
107	.463087	75	35	.000409462	42.5	1.8828	22
108	.487500	40	20	.000231435	43.5	1.9111	12
109	.512195	21	11	.000124619	44.5	1.9397	6
110	.550000	10	6	6.52765E−05	45.5	1.9688	3
111	.555556	5	3	2.96711E−05	46.5	1.9983	1
112	.750000	2	2	1.78027E−05	47.5	2.0283	1
113+	1.000000	1	1	5.93423E−06	48.5	2.0587	0

The last column 7 in Table 5.2 is the product of the probability of your dying in a given year (column 5), the present-value cost correction factor (column 6), and $100 000. This column shows the cost, at your 65th birthday, of insuring you every year thenceforth. The top entry, $x = 65$, is of course the same result as for the one-year term insurance example above.

The total cost of the policy is just the sum of all of the numbers in the last column, $75 791. Is this a reasonable number? The longer you wait before buying a life insurance policy that covers you for the rest of your life, the more it will cost you. Buying the policy at the earlier age of course insures your family

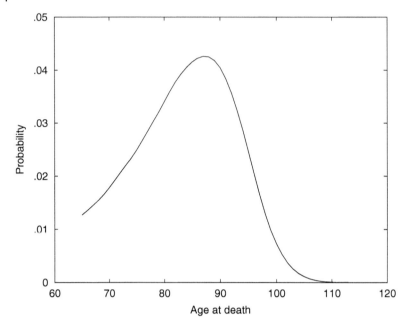

Figure 5.1 Age of death probability starting at age 65.

against your accidental or other untimely death, which is a consideration that just might be more important than the relative costs of the policies.

In practice, it is common to combine the above calculation with some another financial tools: the loan and the annuity. The insurance company will loan you the total amount of money needed and then calculate an annuity, that is, payments for you to make until the loan is repaid in some number of years. This is called Whole Life Policy, it gets fully "paid up" so you don't have to make any further payments to be insured for the rest of your life, and there is a "cash value" of the policy in an account with your name on it that is making the insurance premium payments automatically.

Following through on all of these calculations is beyond the scope of this book; a basic finance or insurance text should be consulted for this.

Social Security – Sooner or Later?

Next, we'll look at a slightly different calculation that will ultimately need the life tables: assume that you have never been married, just to keep the calculations and alternatives simple. Based on your earning history, the Social Security Administration tells you that if you start collecting Social Security payments at the age of 66, you'll get (approximately) $2000 a month for the rest of your life, but if you wait until you're 70 years old to start collecting payments, you'll get (approximately) $2600 for the rest of your life. What should you do?

Let's look at some extreme cases that don't require much calculation first. Assume that you're now just coming up to your 66th birthday and you have to make a decision about when to start collecting Social Security payments. If you're unfortunate enough to be suffering from some terminal disease and know that you probably won't live until you're 70 years old, then your choice is obvious: start collecting the money now and enjoy it while you can. On the other hand, if every member of your family for six generations back has lived well into their nineties and right now you're doing fine health-wise, then you probably should wait until you're 70 years old to start collecting – the higher payments for all of the years past your 70th birthday will more than compensate for the four years' wait.

But what about the intermediate case(s)? You have no terminal disease, your family life expectancy history tends to look like the life tables. What should you do? We'll look at this problem in three different ways. We'll be conservative about interest rates; let's say that 2.5% is a reasonable safe investment (CDs or something like that).

The first way to look at this is to just start taking the money immediately (at age 66) and putting it in the bank. Just to keep the tables small, we'll only compound interest and receive funds twice a year. That is, assume that the Social Security checks are $2000 \times 6 = \$12\,000$ twice a year, and that the bank gives you 1.5% interest on your balance twice a year. The twice-a-year check had you waited until age 70 would have been $2600 \times 6 = \$15\,600$. If you were to take the difference, $\$15\,600 - \$12\,000 = \$3600$ from your savings account twice a year after you reach age 70 and add it to your ($\$12\,000$) check, your financial life wouldn't know which choice you had made – until your savings account ran out.

Table 5.3 shows your savings account balance over the years. The savings account runs out just before your 88th birthday. We won't know what we can conclude from this table until we've gone through the other ways of looking at this problem.

The second way of comparing the two choices is to calculate the present value of both choices at the same date. In this case, a reasonable choice of this date is your 66th birthday, since that's when you have to make a decision. Since you don't know when you're going to die, all you can do is make a list of the present values of each choice versus a possible year of death.

Figure 5.2 shows both present values versus your possible age of death. While the choice of starting payments at age 66 looks better if you die soon, the two curves cross and the choice of starting payments at age 86 looks better if you are to live a long life. The two curves cross just before your 88th birthday. This isn't surprising, because this calculation really is the same as the calculation above; this is just another way of looking at it.

Table 5.3 Social security choice worksheet.

Age	SS check into bank	Withdrawal	Balance
66	12000	—	12000
66.5	12000	—	24180
67	12000	—	36543
67.5	12000	—	49091
68	12000	—	61827
68.5	12000	—	74755
69	12000	—	87876
69.5	12000	—	101194
70	—	3600	99112
70.5	—	3600	96999
71	—	3600	94854
71.5	—	3600	92676
72	—	3600	90467
72.5	—	3600	88224
73	—	3600	85947
73.5	—	3600	83636
74	—	3600	81291
74.5	—	3600	78910
75	—	3600	76494
75.5	—	3600	74041
76	—	3600	71552
76.5	—	3600	69025
77	—	3600	66460
77.5	—	3600	63857
78	—	3600	61215
78.5	—	3600	58533
79	—	3600	55811
79.5	—	3600	53049
80	—	3600	50244
80.5	—	3600	47398
81	—	3600	44509
81.5	—	3600	41577
82	—	3600	38600
82.5	—	3600	35579

Table 5.3 (Continued)

Age	SS check into bank	Withdrawal	Balance
83	—	3600	32513
83.5	—	3600	29401
84	—	3600	26242
84.5	—	3600	23035
85	—	3600	19781
85.5	—	3600	16477
86	—	3600	13125
86.5	—	3600	9721
87	—	3600	6267
87.5	—	3600	2761
88	—	3600	(797)

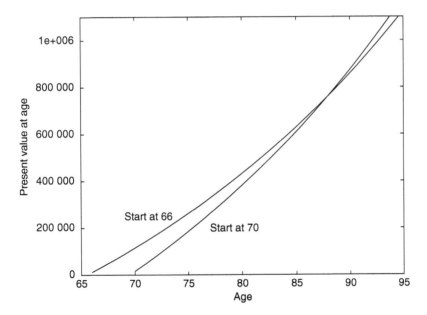

Figure 5.2 Social security present value calculation based on life table.

These calculations give some interesting numbers to look at and ponder, but they don't help to make a decision. You can only make the actual correct decision if you know exactly when you're going to die. Since you don't know this date you have to fall back on the next-best-thing: the probabilities. We'll repeat the present value calculations but we'll weigh them by the information we get from the life tables.

Table 5.4 looks complex, but it's really just a combination of things that have been presented before. The first column is age, starting at age 66. Note that years 73–106 are not shown. However, their contents are included in the calculations.

The second and third columns are the current value of each of the $24 000 payments starting at age 66 and the $31 200 payments starting at age 70, respectively. We are, in this table, assuming that all payments and bank compounding occurs once, annually. This is not as accurate as monthly accruals would be, but it's accurate enough for our purposes. We refer to these as *current* values to avoid confusing them with the previous usage of *present* value – which was the sum of all of the current values up to a given age.

Column 4 is d_x, the number of people dying at age x, taken from the Life Table, Figure 5.1. Column 5 is the PDF of the probability of dying at age x, calculated by normalizing column 4, as described previously.

Column 6 is the probability of seeing your birthday at age x. This is calculated, row by row, by just taking 1 – (column 6, previous row). This column

Table 5.4 Social security choice present value comparison.

Age	CV of $24 k/yr	CV of $31.2 k/yr	d_x	d_x Norm	P of X	24 k Weighted	31.2 k Weighted
66	24 000	—	99 709	.0212	1.0000	24 000	—
67	23 301	—	99 400	.0211	.9788	22 808	—
68	22 622	—	99 369	.0211	.9578	21 667	—
69	21 963	—	99 347	.0211	.9367	20 573	—
70	21 324	27 721	99 330	.0211	.9156	19 524	25 381
71	20 703	26 913	99 315	.0211	.8945	18 519	24 075
72	20 100	26 130	99 302	.0211	.8735	17 556	22 823
...							
...							
107	7143	9286	96 539	.0205	.1424	1017	1322
108	6935	9016	96 355	.0204	.1219	846	1099
109	6733	8753	96 156	.0204	.1015	683	888
110	6537	8498	95 940	.0204	.0811	530	689
111	6347	8250	95 705	.0203	.0607	386	501
112	6162	8010	95 449	.0202	.0404	249	324
113	5982	7777	95 170	.0202	.0202	121	157

starts at 1 because the table starts at age 66 and if you've reached your 66th birthday, it's a certainty that you'll be alive on your 66th birthday (a nonsensical sounding but correct statement).

Column 7 is column 4, the current value of the $24 000 payments, weighted by the probability that you'll be alive to collect this payment. Column 8 is the same for the $31 200 payments. The sum of each of these two columns are the expected values of the two Social Security choices: $388 363 for starting to collect Social Security payments at age 66 and $389 110 for starting to collect Social Security payments at age 70.

These two numbers are so close to each other that there doesn't seem to be a "winner." They are the same to within about ±.1%. Playing around with the interest rate a bit could make them even closer (or further apart). The bottom line seems to be that it really doesn't matter which choice you make. This is actually a very reasonable conclusion and we shouldn't have expected otherwise. The Social Security office accountants have decided that someone in your circumstances (the above assertions) is entitled to Social Security payments with an expected present value of (approximately) $390 000. They then create tables such as Table 5.4 but working from (from our perspective) finish to start rather than start to finish. You pick the way you want to receive the money: you're not "beating the system" and the system is not "beating you" whichever choice you make.

Problems

5.1 A variation of the term life insurance policy is the decreasing term life insurance policy. A decreasing term policy is typically a several year policy, bought as one package, with the policy payoff decreasing by a certain amount every year until it vanishes. A decreasing term policy is typically bought as mortgage insurance or auto loan insurance, i.e. to insure the payoff of a periodically paid loan in case of the death of the policy holder.

Year	Incoming balance	Payment	Insured amount	Age	P (dying)	Premium in year	PV of premiums
1	$40 000	$11 471.67	$28 528.33	30	.001086	$30.98	$30.98
2	$31 381.16	$11 471.67	$19 909.49	31	.001118	$22.26	$20.24
3	$21 900.44	$11 471.67	$10 428.77	32	.001160	$12.10	$10.00
4	$11 471.65	$11 471.67	−$.02				

The table above displays the up-front premium calculation for a five-year term policy. The table shows the amortization table for an auto loan. In order to avoid burying us in loan details, for this loan interest is accrued and payments are made annually, rather than the more common monthly accruals. The annual interest rate is 10%.

Looking at above table, at the beginning of year 1, a $40 000 car is bought by a 30-year-old person and $11 471.67 (a down payment) is paid – this odd number will be explained shortly. The balance owed on the loan is the difference between these two numbers, $28 528.33. A 1-year term policy for this amount must be purchased.

Continuing, at the beginning of year 2, 10% interest is added to the balance, resulting in a new balance of $31 381.16. Again, a payment of $11 471.67 is made, resulting in a balance of $19 909.49 and a 1-year term policy for this amount must be bought. Repeating this procedure for the 3rd year, a $10 428.77 term policy must be bought. The final payment at the beginning of the 4th year reduces the balance to 0,[3] the loan is paid off, and no insurance is needed. The payment amount was calculated for these numbers to work out the result in a zero balance after 4 years.

A Assume that the insurance company is our hypothetical ideal organization: no operating costs, no profits. Also, assume that there is no interest rate on borrowed money. What is the premium cost, payable on day 1 of year 1, for this three-year decreasing term policy?

B Repeat the above assuming that the bank interest rate is 10%.

5.2 Health insurance is different from life insurance in several respects:

1 For many, if not all, policy holders, there will be some payout every year.

2 Most people maintain policies for many years, with payouts growing each year as they age.

3 The payout amount is not a fixed amount as it is with life insurance. There may be a deductible and a cap, but still the payout amount is variable.

This problem will deal with formulating a health insurance policy (again, idealized). The table below shows some of the data gathered about average of peoples' annual medical expenses based upon their ages.

---

3 Since we only calculate payments to the nearest penny, we see that there is a $.02 error. This is typically ignored (in favor of the bank or insurance company).

Age	Avg. annual cost $
25	2000
27	2100
29	2050
31	1900
33	2000
35	2500
37	2300
39	3000
41	3300
43	3400
45	3700

Note: Ages are 2-year groups; i.e. "25" is ages 25 and 26 data averaged together.

From the data, generate a premium amount for every age shown. Create a "5-Year Policy" with equal annual premium payments for 25-, 30-, 35-, and 40-year-olds. Assume a 5% annual interest rate and compound interest annually.

5.3 Based on the life table (Table 5.1), what is the probability of a person on their 60th birthday living until their 70th birthday?

5.4 Two friends are 60 and 65 years old today. What is the probability that at least one of them will be alive ten years from today? (Assume both birthdays are today.)

5.5 This is an extension of Problem 5.4.
I am 60 years old, I have a friend who is 65 years old, and two friends who are 62 years old. Assume all of our birthdays are today. What is the probability that, ten years from today:
A None of us is alive?
B One of us (enumerate) is alive, the other three people are dead?
C My 65-year-old friend and I are alive, both 62-year-old friends are dead?
D Any two of us are alive?

5.6 For this problem, just to keep tables and graphs at a reasonable size, we'll use a reduced Life table as shown here (see Table 5.1 for the complete Life table).

	Nr surviving
Age	To Age x
x	l_x
0	100 000
10	99 266
20	98 974
30	98 081
40	96 796
50	94 354
60	88 751
70	78 123
80	57 418
90	23 879
100	1947
110	10

A Assume an idealized situation – an isolated community where nobody (at least on the average) comes or goes; everybody is born in this community, lives their life here, and dies here. The population is stable. If the birth rate is a constant 100 000 babies every ten years, what is the population?

B What is the (stable) population distribution?

C Assume that all births are from mothers in their 20s and there is an equal number of boys and girls born. What is the average birth rate per mother?

6

The Binomial Theorem

Introduction

Suppose you have ten coins that are numbered 1, 2, ..., 10 and you want to know the probability of flipping a head on the first coin, a tail on the second coin, a ..., etc. This is not a difficult problem. Actually, this isn't *a* problem, it's ten independent problems lumped together in one sentence. The probability of a given result with each coin is .5, so the probability of all ten results is simply $\left(\frac{1}{2}\right)^{10} = \frac{1}{1024} \approx .001$.

But, what if you're interested in a slightly different problem, the problem of getting exactly 6 heads and 4 tails out of 10 flips? You don't care which of the flips gives you heads and which gives you tails so long as when you're done, there are 6 heads and 4 tails. Remember that this is the same as saying you toss ten coins into the air at once to get six heads (and four tails).

One way of solving this problem is to write down every possible result of flipping ten coins and then to count the number of ways that you could get 6 heads and 4 tails. The probability would just be this number of ways multiplied by the probability of getting any one of them, $\left(\frac{1}{1024}\right)$, or equivalently this number of ways divided by 1024. This was the procedure in Chapter 1 when we wanted to see how many ways we could get, say, a total of 6 when rolling a pair of dice (Table 1.2). In the case of the pair of dice there were only 36 combinations, so the writing-down exercise was reasonable. For ten coins there are 1024 combinations, the writing-down exercise impractical. Asking for the probability of getting 15 heads and 5 tails when flipping 20 coins gives us over one million combinations, writing them all down is clearly out of question.

Probably Not: Future Prediction Using Probability and Statistical Inference,
Second Edition. Lawrence N. Dworsky.
© 2019 John Wiley & Sons, Inc. Published 2019 by John Wiley & Sons, Inc.
Companion website: www.wiley.com/go/probablynot2e

The Binomial Probability Formula

To calculate the above probabilities we need the *binomial probability* formula. The problem is *binomial* in the sense that each of the repeated events (the coin flip, the die toss) has two possible outcomes. These outcomes have probabilities p and q, where q is $1 - p$, or the probability of *not p*. As an example of these definitions, if p is the probability of rolling a 5 with a die, $p = \frac{1}{6}$, then q is the probability of rolling anything but a 5 with the die $q = 1 - \frac{1}{6} = \frac{5}{6}$.

We flip a coin n times and want to know the probability of getting k heads. We know that $0 \le k \le n$, which is a fancy way of saying that k can get as small as 0 (no heads) or as large as n (all heads). The probability of getting k heads in k flips is p^k.

The probability of getting everything else tails is then Prob of "*the rest of them*" *tails* = q^{n-k}.

Taking this last expression one piece at a time, remember that if the probability of getting a head is p, then the probability of getting a tail is q. This seems trivial for a coin when $p = q = .5$, but if we had a "funny" coin that lands on heads 90% of the time, then $p = .9$ and $q = .1$. If we flip n coins and want k of them to be heads, then $n - k$ of them must be tails.

The probability of getting k heads out of n coin flips when the probability of a head is p is therefore (*number of ways*) $p^k q^{(n-k)}$, where "number of ways" means the number of ways that we can achieve k heads when we flip n coins.

Before looking into a formula for calculating "number of ways," consider a couple of examples based on situations where it's easy to count the "number of ways." In situations where you can easily count the number of ways, you don't need the whole binomial probability formula in the first place, but just to solidify the use of the terms in the formula, we'll go through a couple of these cases.

Example 6.1 What is the probability of getting exactly 2 heads from 5 coin flips? In this example $n = 5$ (the number of coins), $k = 2$ (the number of heads), and $p = .5$. Therefore,

$$p^k q^{n-k} = (.5)^2 (.5)^3 = (.25)(.125) \approx .0313$$

Table 6.1 shows all (32) possible combinations of flipping 5 coins. In the table, there are 2 heads in 10 (of the 32) rows (bold faced rows). The probability of getting exactly 2 heads when flipping 5 coins is therefore $(10)(.0313) \approx .313$.

The probability of getting exactly 1 head when flipping 5 coins is (5) $(.0313) \approx .157$, and so on.

Table 6.1 Five coin flip possibilities, two-heads results highlighted.

Flip 1	Flip 2	Flip 3	Flip 4	Flip 5
H	H	H	H	H
H	H	H	H	T
H	H	H	T	H
H	H	H	T	T
H	H	T	H	H
H	H	T	H	T
H	H	T	T	H
H	**H**	**T**	**T**	**T**
H	T	H	H	H
H	T	H	H	T
H	T	H	T	H
H	**T**	**H**	**T**	**T**
H	T	T	H	H
H	**T**	**T**	**H**	**T**
H	**T**	**T**	**T**	**H**
H	T	T	T	T
T	H	H	H	H
T	H	H	H	T
T	H	H	T	H
T	**H**	**H**	**T**	**T**
T	H	T	H	H
T	**H**	**T**	**H**	**T**
T	**H**	**T**	**T**	**H**
T	H	T	T	T
T	T	H	H	H
T	**T**	**H**	**H**	**T**
T	**T**	**H**	**T**	**H**
T	T	H	T	T
T	**T**	**T**	**H**	**H**
T	T	T	H	T
T	T	T	T	H
T	T	T	T	T

In order to derive a formula rather than counting rows in possibly very long tables, we must introduce the topic of permutations and combinations of numbers.

Suppose we have 4 objects, labeled A, B, C, and D, and we want to see how many ways we can order them, such as ABCD, ADCB, GDCA, etc.

For the first letter, we have 4 choices (A, B, C, or D). For the second letter, we only have 3 choices: we already used one of the choices for the first letter. For the third letter we have 2 choices, and for the last letter we have only 1 choice left. This means we have (4)(3)(2)(1) = 24 choices.

The calculation (4)(3)(2)(1) is called *four factorial*, the notation for which is 4!. In general, the number of ways that we can order n objects is n!. Spreadsheet programs and scientific calculators have a factorial function available, so this calculation isn't necessarily tedious. However, it does get out of hand very quickly: 4! = 24, 5! = 120, 6! = 720, etc. From this short list you might have spotted a fundamental principle of factorials: $n! = n[(n-1)!]$. This is another example of a *recursive* relationship.

Permutations and Combinations

What if we want to order 6 items, but only look at, say, 4 of them at a time? In other words, if we have a box with 6 items in it, how many ways can we order any 4 of them? The solution to this is an extension of the factorial calculation: For the first item we have a choice of 6, for the second item a choice of 5, for the third item a choice of 4, and for the fourth item a choice of 3. Putting all of this together, the number of ways I can order 6 items, 3 at a time is (6)(5)(4)(3).

This looks like a "piece" of a factorial calculation. We can write (6)(5)(4)(3) as

$$(6)(5)(4)(3) = \frac{(6)(5)(4)(3)(2)(1)}{(2)(1)} = \frac{6!}{2!} \tag{6.1}$$

Therefore, we can write the number of ways we can order n items, k at a time, as

$$P_{n,k} = \frac{n!}{(n-k)!} \tag{6.2}$$

Note the new notation here. This formula is known as the *permutation* of n things taken k at a time.

Now, there is a slight difference between $P_{n,k}$, the *number of ways we can order n things k at a time* and the *number of ways a given goal can be accomplished*. The difference is a consequence of the fact that in the former case order counts (ABCD is not the same as BACD), whereas in the latter case order does *not* count (ABCD is the same as BACD, is the same as DCAB, etc.). When you want the number of ways you can get 4 heads out of 6 flips of a coin, you don't care if you get HHHHTT or TTHHHH or THHHHT or... . The permutation formula is predicting too large a number! A good example of this is a dealt hand of cards. Your hand is totally determined by the number of cards you end up with, not the order in which you dealt them.

The number of ways we can order k items is $k!$. Therefore, if we take $P_{n,k}$ and divide it by $k!$, we'll have reduced the number of ways of ordering n items k at a time by the number of ways we can (re)arrange the k items. This is just what we are looking for: the number of ways our goal can be accomplished.

This relationship is known as the *combination* of n things taken k at a time, and is written

$$C_{n,k} = \frac{P_{n,k}}{k!} = \frac{n!}{(n-k)!k!} \tag{6.3}$$

Sometimes $C_{n,k}$ is written $\binom{n}{k}$ or $C\binom{n}{k}$; these are different notations for the same calculation.

$C_{n,k}$ is easy to calculate, by hand for n small and using spreadsheets or scientific calculators for larger values of n. As an example, the combination of 10 things taken 6 at a time is

$$C_{10,6} = \frac{P_{10,6}}{4!} = \frac{(10)(9)(8)(7)}{(4)(3)(2)(1)} = 210 \tag{6.4}$$

Putting all of this together, the probability of getting 6 heads out of 10 flips of a coin (or a flip of 10 coins) is

$$C_{10,6}(.5)^6(.5)^4 = (210)(.015625)(.0625) \approx .205$$

What about the probability of getting 4 heads out of 10 flips? Since the probability of a single head = the probability of a single tail = .5, then the probability of getting 4 heads out of 10 flips should be the same as the probability of getting 4 tails out of 10 flips, which of course is the same as the probability of getting 6 heads out of 10 flips.

If you want the probability of getting *at least* 6 heads when you flip 10 coins, you must add up all of the individual probabilities: the probability of getting 6 heads (out of 10 flips), 7 heads, 8 heads, 9 heads, and 10 heads when you flip 10 coins.

Figure 6.1 shows the probability of getting k heads out of 10 flips. k can take on values from 0 to 10 ($0 \le k \le 10$). These 11 probabilities add up to 1, as they should, Figure 6.1 is a discrete probability distribution function (PDF).

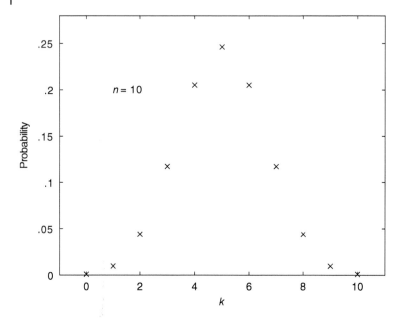

Figure 6.1 Binomial probabilities for ten (fair) coin flips.

Large Number Approximations

Figure 6.1 looks suspiciously like (points on) a normal curve. The mean is 5. This makes sense: if we flip 10 coins, each with a probability of .5 of heads, then we would expect the mean to be (10)(.5) = 5.0. Remember that, for a normal distribution, the mean is the same as the expected value. In general, we should expect the mean to be the product of the number of coin flips (or whatever we're doing), n, and the probability of success at each even, p,

$$\mu = np \tag{6.5}$$

Figure 6.2 shows a normal distribution with the mean as shown above and the standard deviation equal to[1]

$$\sigma = \sqrt{np(1-p)} \tag{6.6}$$

overlaid on the discrete binomial formula data (Figure 6.1).

The normal distribution is an excellent approximation to the binomial distribution for any but small values of n (say, $n < 5$), and the approximation gets better as n gets larger, so long as p isn't very small or very large. This shouldn't

---

1 The derivation of Eq. (6.6) is beyond the scope of this book.

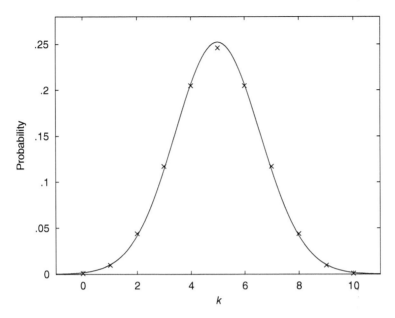

Figure 6.2 Figure 6.1 with fitted normal distribution.

be a surprise, because in Chapter 3 we showed that every distribution starts looking normal when you add more and more cases of it (the distribution) together.

Using the above formulas for the mean and the standard deviation of a normal distribution approximation to the binomial distribution can save a lot of time when looking at problems with large values of n. It is necessary to be able to calculate values of the normal distribution; but again, this is available on spreadsheets and scientific calculators (as is the binomial distribution itself).

Figure 6.3 shows a case where the binomial and normal distributions look similar, but they're not an exact overlay. The numbers that were used to generate Figure 6.3 are $n = 1$ billion ($= 1 \times 10^9$ in scientific notation) and $p = 10/(1 \text{ billion}) = .00\,000\,001$ ($= 1 \times 10^{-8}$ in scientific notation). The expected value is $np = 10$. This is a pretty unlikely event; only 10 "hits" out of 1 billion tries! Since the normal PDF looks somewhat like the binomial PDF, the standard deviation calculated using the above formula is reasonable. Using this formula,

$$\sigma = \sqrt{np(1-p)} = \sqrt{\mu(1-p)} = \sqrt{10(.9999999)} \approx \sqrt{10} \approx 3.16 \qquad (6.7)$$

The 95% confidence interval of a normal distribution is $\pm 2\sigma$, which in this case is 6.32 above and 6.32 below the mean. However, the binomial

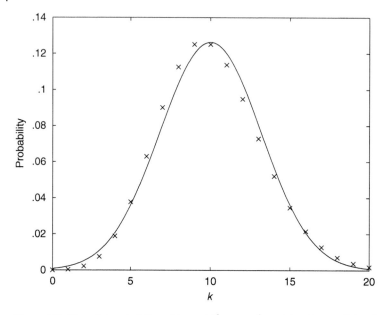

Figure 6.3 Binomial probability with $n = 10^9$, $p = 10^{-8}$, and fitted normal distribution.

distribution is a discrete distribution in which the number of successes, or "hits" must be an integer. What does 6.32 mean in this situation?

The normal distribution is a continuous function approximation to the binomial distribution. The binomial distribution is a discrete PDF: it is only defined for integer values of k. If we want to take advantage of the normal PDF approximation, we'll just say that the mean ±6 is a confidence interval of "just under" 95% and leave well enough alone.

The Poisson Distribution

Consider a *really* improbable event: $n =$ a billion (1×10^9), but $p =$ one in a billion (1×10^{-9}). The binomial and normal PDFs for these values are shown in Figure 6.4.[2]

Figure 6.4 shows that in the extreme case of a very rare occurrence out of very, very many trials, the binomial distribution doesn't look at all like the

---

2 These values are tricky to calculate by hand. Approximations using the "binomial expansion" which is related to the binomial probability formula are necessary. Interestingly, for n large, $p = 1/n$, and $k = 0$, 1, 2 the value of n doesn't matter.

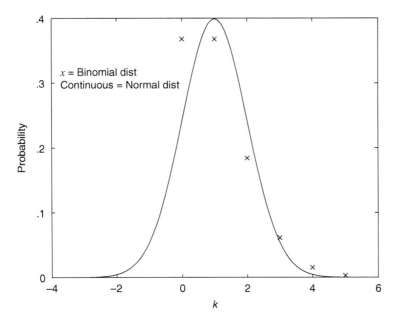

Figure 6.4 Binomial probability with $n = 10^9$, $p = 10^{-9}$, and fitted normal distribution.

normal distribution. In this example, the normal distribution would extend its 95% confidence interval into negative values of k, which makes no sense whatsoever.

The Poisson distribution is a discrete distribution which is defined only for non-negative integer values of k. The theory behind the Poisson distribution is beyond the scope of this book. The important properties of the Poisson distribution are:

1) The Poisson distribution is a discrete distribution, valid only for integer arguments.
2) When the Poisson distribution is approximating a binomial distribution, the mean value (usually called λ) is $\lambda = np$.

$$\lambda = np \tag{6.8}$$

3) The standard deviation is

$$\sigma = \sqrt{\lambda} \tag{6.9}$$

Formally, the value of the Poisson distribution for k occurrences with a mean value of λ is

$$P_{k,\lambda} = \frac{e^{-\lambda}\lambda^k}{k!} \tag{6.10}$$

This formula won't be directly used very often, if at all, but is included here for completeness.

The Poisson distribution is a good approximation of the binomial distribution for $n > {\sim}200$ and either p or q very close to 1 (which of course makes q or p, respectively, very close to 0). This distinguishes it from the normal distribution, which, as was shown above, is a poor choice as an approximation is when p is close to 0 or 1.

In addition to its value as an approximation to binomial probability, the Poisson distribution describes the probability of k occurrences of an event if the events occur at a known rate (λ) and the occurrence of each event is independent of the occurrence of the last event.

Figure 6.5 shows Poisson distributions for several values of λ.

As an example of using the Poisson distribution, suppose that every day you take a walk through the park at lunchtime. You walk down a quiet lane and count the number of cars that pass by you. Over the course of a year the average number of cars that pass by you is 10 per hour. The Poisson distribution (Figure 6.5, $\lambda = 10$) is the PDF for the number of cars that pass you each day.

In Figure 6.5, $\lambda = 10$ looks symmetric and not unlike a normal distribution. However, note that the mean occurs at $k = 10$, while the value of the PDF at

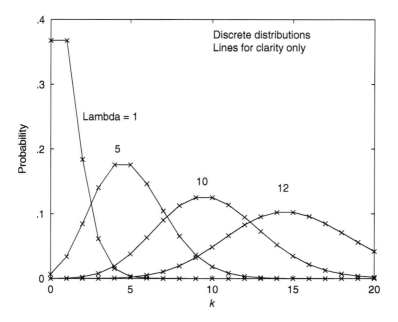

Figure 6.5 Poisson distribution, several cases.

$k = 9$ is the same as the value at the mean. This PDF is asymmetric. However, for $k > 5$, the 95% confidence interval may be approximated reasonably well by assuming the PDF is normal.

When $\lambda = 10$, $\sigma = \sqrt{\lambda} = 3.6$, so the normal curve approximation for the 95% confidence interval is 14.4, centered at 10. Using the exact Poisson distribution data, we must approximate with the nearest points. Figure 6.6 shows the cumulative distribution function (CDF) for this distribution.

From the CDF, the closest points we can find for calculating the 95% confidence factor are CDF = .973 at $k = 17$ and CDF = .029 at $k = 12$. We have a 94.4% confidence factor from $k = 5$ to $k = 17$. This is close enough to 95% that the difference is rarely significant.

Note that the confidence interval is not symmetric about the mean ($k = 10$). Asymmetric PDFs open up a whole can of worms about defining confidence intervals. There is no limit to how many definitions we can come up with. Since these tend to be specific to the PDF at hand, we'll skip getting involved in this discussion.

For smaller values of k than 10, the Poisson PDF is notably asymmetric and you must go to the formulas (spreadsheet, scientific or financial calculator) for confidence interval calculations.

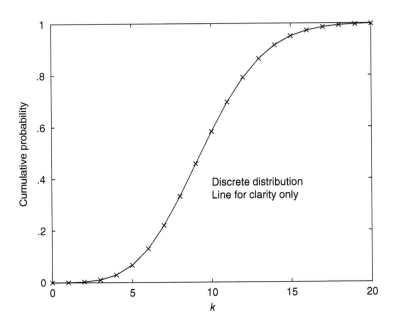

Figure 6.6 CDF of Poisson distribution, $\lambda = 10$.

Disease Clusters

When people with some tie between them such as family relations, friendships, residence neighborhood, drug (legal or illegal) usage, and such start developing unusual diseases, there's reason to be suspicious. If the average incidence in the United States of some particular disease is 1 in 100 000 per year, then the expected incidence in a city of 1 million population of this disease is 10 per year. If the incidence in our city (of 1 million people) is 15 this year, we would work through the probabilities and conclude that we should keep our eye on things but not jump to any conclusions. However, if 6 of these 15 know each other, or are related to each other, or all live within ¼ mile of the Acme Chemical Plant, or something like that, however, then it's not a coincidence. Or is it?

Historically, important public health discoveries arose from studies that began with observations such as the above. Two examples are the long-term danger of asbestos inhalation and the initial understanding of the nature of AIDS. Unfortunately, things sometimes go the wrong way. The courts decided that silicone breast implants caused serious illnesses and awarded enough money to plaintiffs that the implant manufacturer, Dow Chemical, declared bankruptcy because it couldn't pay all the settlements. Several large subsequent studies showed that there was no evidence supporting the tie between these breast implants and the diseases, but we can be sure that this information did not cause any guilt-ridden plaintiffs to mail the money back.

We want to consider the inevitability of a highly unlikely event given enough opportunities for the event to occur. Here are two examples:

First: Probability of an unlikely event in a large population. "Deadly cosmic rays are randomly hitting the earth and striking 100 000 people nightly." What should the incidence rate (expected value and standard deviation) be in your city of 1 million inhabitants?

Second: We have a disease (or cosmic rays as above) that affect one person in 100 000, with all the statistics attached to this. We have a new factor (watching too much reality TV?) that increases your chance of contracting the disease or succumbing to the cosmic rays, or …, that we think is causing 5 extra cases of the disease per 100 000. What should be the expected number of cases and the standard deviation of this?

Clusters

Suppose you live in a city of ¼ million people. If there's a non-communicable disease (we'll look at contagion in a later chapter) such as a brain cancer that afflicts, say, 1 person in 5000 each year, then we would expect the average incidence of this disease in your city to be (using the binomial probability formula)

$$\mu = np = (250\,000)(.0002) = 50 \qquad (6.11)$$

cases per year.

Again, using the binomial probability formula, the standard deviation should be

$$\sigma = \sqrt{\mu(1-p)} = \sqrt{50(1-.0002)} = 7.07 \qquad (6.12)$$

As an aside, note that, to a very good approximation,

$$\sigma = \sqrt{\mu(1-p)} = \sqrt{50(1-.0002)} \approx \sqrt{50} = \sqrt{\mu}, \qquad (6.13)$$

which is the relationship we see when dealing with a Poisson distribution. In this case the normal distribution is also a good approximation. Without showing the graph, we'll (re)assert that the binomial formula for the probability of some number of case overlaps the normal distribution curve for the same μ and σ so well that it's hard to tell that two separate curves have been drawn.

The 95% confidence interval is therefore essentially between 43 and 57 cases. Any number of cases falling outside these numbers is typically deemed to be "statistically significant."

Assume that this year your town has 70 cases of this disease. This is certainly well above the expected statistical variation and deserves further study.

On the other hand, having only 30 cases this year would also deserve further study, and newspaper articles, and television specials discussing it – maybe there's something very beneficial going on in this town with regard to incidence of this disease. Somehow this never seems to happen.

From the binomial formulas, the probability of getting 70 cases this year is just .0010. The probability of getting 70 or more cases is .0026. Now, let's assume that there are 1000 towns around the country more or less like yours. Using these same formulas yet one more time, we find that the expected number of towns with 70 or more cases of this disease is approximately 2.5. It's pretty certain that there will be at least one town out there with 70 or more cases of this disease, even though when you look at just one of these towns at a time, it's statistically very unlikely that this can happen unless there's a special cause.

The bottom line conclusion would still be "Be Very Careful About Assigning A Cause To What Might Be a Truly Random Event." Look for asbestos, look for toxic industrial waste, look for nuclear materials, but also be open to just having plain old bad luck.

Looking forward, Chapter 16 will more formally treat frequentist statistical inference and how to look at some sample data. Chapter 11 will introduce Bayesian statistics, a different way to think about probabilities and learning more as we accumulate data.

Problems

6.1 A drawer contains two pairs of socks – one white and one brown. Two socks are drawn at random. What is the probability that they are a matched pair? Do not create a table, use the appropriate permutation/combination formulas.

6.2 An ordinary playing card deck has 52 cards, 13 of each suite (Hearts, Diamonds, Clubs, Spades) ranging from 1 (the ace) to 13 (the king). We draw two cards. What is the probability that they are both spades?

6.3 Using an ordinary playing deck, what is the probability of drawing 5 cards and getting all spades?

6.4 You are about to start on a trip that takes between 1 and 2 hours, depending on traffic and the weather. Historically, the distribution of traveling times is uniform. You want to arrive at a particular time, plus or minus a minute. How many times must you take this trip to assure that this happens at least once?

6.5 What would the answer to Problem 6.4 be if we said *exactly one success* rather than *at least one success*?

6.6 A fair coin is flipped 15 times.
   A What is the probability of getting more heads than tails? Use the binomial probability formula for your calculations.
   B Repeat the above using the normal distribution approximation of the binomial probability formula. How accurate is your result?

6.7 Repeat Problem 6.6, but for a fair coin flipped 14 times.

6.8 Assume that the use of the letters in the English alphabet is uniformly distributed in a typical book (not true). What is the probability of a chosen letter appearing between 4% and 7% of the time on a page? Make any reasonable assumptions necessary.

6.9 Repeat the approximation part of Problem 6.8, but use a Poisson distribution rather than a normal distribution as the approximation function.

6.10 A housing development has 2000 homes, each with one kitchen sink. The probability of the sink having a leaky faucet is 1/5000. What is the probability that 0, 1, 2, or 3 homes in the development have leaky faucets? Use the binomial probability formula, the Poisson and normal approximations, and comment on the relative success of the approximations.

7

Pseudorandom Numbers and Monte Carlo Simulations

Random Numbers and Simulations

Suppose we want to generate a list of uniformly distributed random numbers. These could be integers between 0 and 10, or real numbers between .0 and 1.0, or some other requirement. We want to do this because it allows us to write computer programs that perform *Monte Carlo* Simulations. A Monte Carlo simulation lets us perform a simulation of any situation involving random numbers that we can describe clearly, and look at possible results. If we repeat this process many times, then we get to see the statistical variations in these results. As a simple example, let's look at our simple coin flip again. If we had a list of randomly generated zeros and ones, then we could assign the zero to a coin flip heads and a one to a coin flip tails. By counting the number of zeros and ones and looking at the ratio, we could conclude that the probabilities of heads and tails is .5 each. If we looked at, say 1 million lists of zeros and ones, each list containing 100s of these numbers, then we could calculate the standard deviation we'd expect in a 100 coin flip game. While in this simple scenario we know how to calculate all of these things and don't need a simulation, there are an infinite number of complicated situations where we cannot calculate and predict the results – and then a simulation is a powerful tool.

Pseudorandom Numbers

Hardware random number generators are machines that measure random noise or physical phenomena that we believe to be truly random (with some known PDF) and convert these measurements to strings of numbers. However, these machines are costly and run slowly (here we are defining slowly as compared to the blazing calculating speeds of even small personal computers). We prefer to use something inexpensive and fast.

Probably Not: Future Prediction Using Probability and Statistical Inference,
Second Edition. Lawrence N. Dworsky.
© 2019 John Wiley & Sons, Inc. Published 2019 by John Wiley & Sons, Inc.
Companion website: www.wiley.com/go/probablynot2e

Computer scientists and mathematicians have learned how to generate what are called *pseudorandom* numbers using computer programs called, reasonably enough, *pseudorandom number generators* (PRNGs).

A computer program that is fed the same input data will produce the same results every time it is run. Numbers coming out of a PRNG are therefore not random. Every time you start a PRNG with the same input parameters (call *seeds*), it will produce the same string of numbers. This sequence of numbers will successfully approximate a sequence of random numbers based on how well it achieves a sophisticated set of criteria. These criteria include the probabilities of successive digits occurring, the mean and sigma of a large sequence, whether some properties of the generator can be deduced from studying a string of numbers that this generator produced, etc.

The mathematics of pseudorandom number sequences and trying to refute the randomness of a proposed sequence is a very sophisticated field of mathematics. The following is merely a surface-scratching introduction.

PRNGs must depend on a fixed formula, or algorithm, to generate their results. Sooner or later every PRNG will repeat. Therefore, another requirement for a good PRNG is that it doesn't repeat until it has generated many more numbers than we need for our problem(s) at hand.

As a very simple example, for a string of pseudorandom integers between 1 and 10 consider the list 1, 2, 3, 4, 5, 6, 7, 8, 9, 10, 1, 2, 3, 4, 5, 6, 7, 8, 9, 10, 1, 2, 3, 4,.....

This list looks pretty good if we calculate the mean and the standard deviation. However, since this list repeats after every tenth number, it will only be usable – if it's usable at all – when we need ten or less numbers. Using only the first ten numbers gives us such a short list that statistics tend to be poor, but let's continue just to demonstrate the evaluation criteria. We would immediately flunk even the short list of the first ten numbers from the above list because, after the first few numbers, it's pretty easy to guess the next number. This is easily circumvented – we'll just scramble the list to, say, 4, 7, 3, 5, 8, 2, 10, 1, 6, 9.

Even after scrambling, this list is still suspicious – there are no repeats. While a sequence of 10 real numbers between 1 and 10 would almost never have any repeats, a list of integers in this range would almost always have at least one repeat. Later in this chapter we'll see how to use a Monte Carlo simulation to calculate the probability of this happening.

The Middle Square PRNG

An early computer-based PRNG is known as the *middle square* method. Suppose we want a random (real) number between 0 and 1, with 4 digits of resolution (how to extend this to any number of digits will be obvious).

Table 7.1 Middle square PRNG example.

Seed	Seed squared	Extraction	Pseudorandom NR
3456	11943936	9439	.9439
9439	89094721	0947	.0947
947	896809	8968	.8968
8968	80425024	4250	.4250
4250	18062500	0625	.0625
625	39065	3906	.3906
3906	15256836	2568	.2568
2568	6594626	5946	.5946
5946	35354916	3549	.3549
3549	12595401	5954	.5954
5954	35450116	4501	.4501
4501	20259001	2590	.2590
2590	6708100	7081	.7081
7801	50140561	1405	.1405
1405	1974025	9740	.9740

Start with a 4-digit "seed" number as shown in Table 7.1. We'll start with 3456 as a seed, as seen in the upper left corner of the table. When we square a 4-digit number, we get an 8-digit number – the second column in the table. Extract the inner 4 digits of the 8-digit number (column 3) and divide it by 10000 (column 4) to get a number between 0 and 1. Finally, use these new (inner) 4 digits as the seed for the next random number (row 2, column 1) and repeat the process.

Table 7.1 shows only the first 15 numbers of the sequence. How does it do as a pseudorandom distribution? A uniform distribution between 0 and 1 should have a mean of .500. These 15 numbers have a mean of .476. Sigma should be .287, this group has a sigma of .308. The smallest number is .0625, the largest number is .974. So far, this group looks pretty good.

Unfortunately, this procedure has some pathological problems. Table 7.2 shows what happens if the number 3317 comes up. The sequence decays to zero and never leaves it again. Table 7.3 shows what happens if the number 1600 shows up. In this case the sequence gets *stuck* in a pattern that repeats every fourth iteration. This doesn't look very random at all.

This procedure could be "patched" by having the computer program that's generating these numbers look for these pathological situations and jump away

Table 7.2 Pathological decaying middle square PRNG example.

Seed	Seed squared	Extraction	Pseudorandom NR
3317	11 002 489	0024	.0024
24	576	0005	.0005
5	24	0	.0000
0			

Table 7.3 Pathological repeating middle square PRNG example.

Seed	Seed squared	Extraction	Pseudorandom NR
1600	2 560 000	5600	.5600
5600	31 360 000	3600	.3600
3600	12 960 000	9600	.9600
9600	92 160 000	1600	.1600
1600			

from them by perturbing the seed, but then we'd have to carefully examine the consequences of these fixes to the statistics of the numbers generated. In general, this is really not a good way to proceed.

The Linear Congruential PRNG

A more modern (and successful) PRNG procedure is the *linear congruential* generator. The formula is very simple. There are 3 constants: A, B, and M. With a starting value V_j, the next value, V_{j+1} is

$$V_{j+1} = (AV_j + B) \bmod M \tag{7.1}$$

mod is shorthand for a modulus operation; take the number in front of mod, divide it by M, and only look at the remainder. A few examples of this are $5 \bmod 4 = 1$, $7 \bmod 5 = 2$, and $6 \bmod 3 = 0$.

Table 7.4 shows this formula used for the values $A = 11$, $B = 17$, and $M = 23$.

The starting value (the "seed") is 4. Since this formula will generate integers between 0 and 22, a random number between 0 and 1 is generated by dividing the results of this formula by M.

A continuous uniform distribution of random numbers between 0 and 1 would have a mean of .500 and a sigma of .289. The list in Table 7.4 has a mean

Table 7.4 Linear congruential PRNG $A = 11, B = 17, M = 23$.

Index	V_j	V_{j+1}	Nr
0	4	5	.217
1	5	16	.696
2	16	22	.957
3	22	19	.826
4	19	9	.391
5	9	14	.609
6	14	0	.000
7	0	7	.304
8	7	15	.652
9	15	11	.478
10	11	13	.565
11	13	12	.522
12	12	1	.043
13	1	18	.783
14	18	21	.913
15	21	8	.348
16	8	3	.130
17	3	17	.739
18	17	10	.435
19	10	2	.087
20	2	6	.261
21	6	4	.174
22	4	5	.217

of .450 and a sigma of .287. The lowest number in the list is 0, the highest number is .957. There is no (or certainly no obvious) pattern to the spacing between the numbers, or any other discernible characteristic. This looks like a good pseudorandom list.

The *linear congruential* generation of pseudorandom sequences will repeat *at most* after *m* numbers, as Table 7.4 illustrates. Whether or not it actually makes it to M numbers before repeating depends on the choices for A and B. Common software today uses an algorithm with M greater than 10 million, in some cases up into billions, and appropriately chosen values for A and B.

Once we have a PRNG that gives us uniformly distributed numbers between 0 and 1, we can easily extend this to a PRNG that gives us uniformly distributed numbers between any lower and upper limits. If V_1 is such a number between 0 and 1, and we want numbers between a lower limit a and an upper limit b, then V_2 given by

$$V_2 = a + (b - a)V_1 \qquad (7.2)$$

Next, we'd like to look at the generating pseudorandom sequences with other than uniform distributions. There are many ways of doing this and we're only going to discuss two of them. These two aren't optimum in any sense, but they work and they give good insight into the processes involved. There are more modern, more efficient ways to do both jobs. Both of these procedures assume the availability of a PRNG such as described above, a PRNG that generates a uniform (pseudo) random sequence between 0 and 1.

A Normal Distribution Generator

The first example is a normal distribution generator for distributions with parameters μ and σ. We'll take advantage of the Central Limit Theorem; we know that adding up random numbers from any distribution will ultimately give us a normal distribution. Our choice here for *any* distribution is the uniform distribution between 0 and 1.

Figure 7.1 shows this uniform distribution, labeled $n = 1$. This is followed by graphs of several of the distributions we get when we add this uniform distribution to another uniform distribution, then add the results to another uniform distribution, and so on – exactly as was described in Chapter 3. We keep going until we've combined 12 such distributions ($n = 12$ in the figure). For this figure, each distribution has been scaled so that its peak is 1. This isn't necessary, but it makes the composite graph of these figures easier to examine.

Combining a distribution from 0 to 1 with another distribution from 0 to 1 results in a distribution from 0 to 2, combining this with a distribution from 0 to 1 results in a distribution from 0 to 3, and so on. Since these are symmetric distributions, the mean occurs exactly in the middle. Therefore, the distribution from 0 to 12 ($n = 12$ case) has a mean of 6.

Twelve distributions were added because the standard deviation of this sum of 12 uniform distributions is exactly 1. This is very convenient because it leads to the following procedure for generating normally distributed random variables:

1) Generate 12 uniformly distributed random variables between 0 and 1, x_i

2) Add them up: $\sum_{i=1}^{12} x_i$

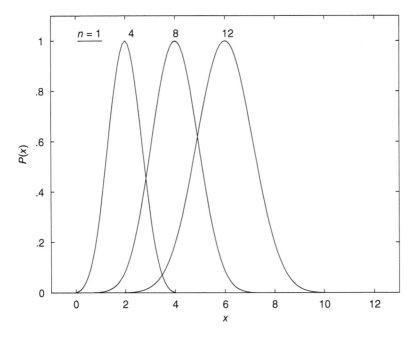

Figure 7.1 Uniform distribution sums for normal distribution simulation.

3) Subtract 6: $\displaystyle\sum_{i=1}^{12} x_i - 6$ This will bring the mean to 0 without changing the standard deviation

4) Multiply by the desired σ: $\displaystyle\sigma\left(\sum_{i=1}^{12} x_i - 6\right)$

5) Add the desired mean: $\displaystyle\sigma\left(\sum_{i=1}^{12} x_i - 6\right) + \mu$

The results of this procedure are of course only as good as the PRNG used. Also, it requires using the PRNG 12 times for every normally distributed number generated. Still, it works well and it is simple and clear.

An Arbitrary Distribution Generator

The following procedure will generate random variables for an arbitrary PDF so long as the x values are only allowed over a finite range. That is, we must be able to specify specific starting and stopping values for x.

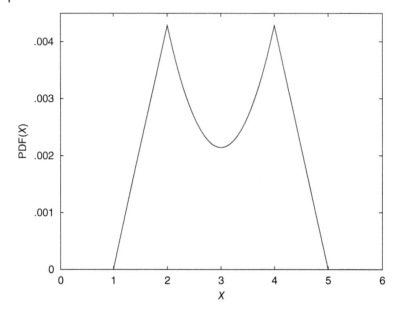

Figure 7.2 Arbitrary function chosen for random number generation example.

Figure 7.2 shows an arbitrary PDF with $1 \leq x \leq 5$. Now,

1) Scale this PDF so that the maximum value is 1. Figure 7.3 shows the result of the scaling.
   Choose a random number between 1 and 5 using a uniform distribution random variable generator. Then, choose a second random number between 0 and 1.
   Consider Figure 7.3.
2) Let the first number chosen be your trial value of x. If the second number chosen is greater than the value of the distribution at the trial value of x, go back to step 2 and start over.
3) If the second number chosen is less than the value of the distribution at the trial value of x, then the trial value of x is the "output" of the procedure.

In Figure 7.3, two examples of this procedure are shown: $x = 3.2$ is a (randomly chosen) trial value of x; .81 is the second random number. The point (3.2, .81) is above the distribution function curve, so these numbers are rejected and the procedure restarted. $x = 2.5$ is the second trial value of x, .4 is the second random number. The point (2.5, .4) is below the distribution function curve, so $x = 2.5$ is the output of this random number generator.

If we created a long sequence of random numbers using this procedure, we would expect to see many more numbers from the regions where the distribution function is large (close to 1) than I would from where it is small. Fortunately, this is exactly what we're looking for.

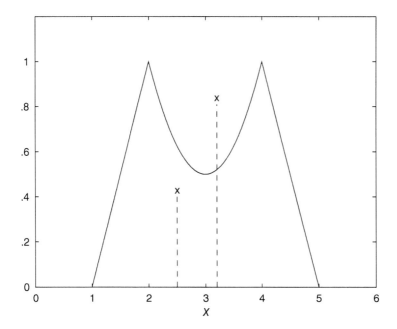

Figure 7.3 Two choices of *X* for demonstration of random number generation procedure.

This procedure is not very efficient in that it's not very *smart*. It doesn't really calculate anything, it just keeps trying and rejecting numbers that don't fit the criterion. It is very good in that it's robust (it never gets into numerical trouble) and it can be explained with absolutely no equations!

There are other, more efficient, procedures that are more formal mathematical calculations, but we now have the tools needed to do what we want to do.

Monte Carlo Simulations

The ability to generate (pseudo) random numbers satisfying any desired distribution function enables us to create arbitrary Monte Carlo simulations. This is a very powerful tool for studying processes that are too complicated for formal analysis.

As you will soon see, the Monte Carlo method is a brute force method. It was once likened to "putting a paper bag over your head and running down the road." This isn't far from the truth. The following Monte Carlo Simulations will make this point clear. The first simulation will be a simple mathematical "game," the second a hypothetical real world situation.

The first simulation will be to generate a list of ten random integers between 0 and 9. Then the computer will count the number of repeated integers in the list and save this number. We'll repeat this procedure many times, making

Table 7.5 Lists of 10 digit integer strings for Monte Carlo simulation.

Nr										
1	5	0	4	5	7	4	9	9	4	7
2	1	3	4	9	7	3	3	3	0	9
3	4	7	1	7	10	6	1	3	0	10
4	3	0	8	8	4	1	1	8	6	3
5	3	2	6	7	7	9	7	10	4	2
6	5	7	4	9	10	4	0	10	10	9
7	5	3	2	10	5	10	9	5	7	8
8	6	2	3	0	7	9	0	9	0	4
9	6	8	2	5	3	6	9	3	2	8
10	7	9	8	5	9	2	9	6	7	10
11	5	2	2	4	6	7	6	1	2	10
12	4	5	10	10	10	6	2	7	3	7
13	5	4	9	6	3	0	9	4	9	10
14	3	4	8	2	8	4	1	7	8	0
15	2	2	5	2	7	2	7	0	10	1
16	3	5	10	9	2	0	5	8	1	0
17	3	3	6	6	4	7	3	10	1	0
18	7	4	4	7	5	2	6	2	10	2
19	5	7	1	3	2	1	7	7	5	8
20	1	6	3	7	6	9	10	0	0	3

sure that we never restart the PRNG at its initial seed values. Then we'll summarize the results.

Table 7.5 shows the first 20 lists of 10 single-digit integers that the computer program produced. None of these lists has no repeats. The first list has repeats of 4, 5, and 7 (shown in italics).

Running the program 1000 times yielded no lists have no repeats. After 5000 iterations, the program found 1 list that had no repeats. That's 99.98% of the lists that had at least one repeat. After 10 000 iterations this was 99.96%, after 100 000 iterations this was 99.956%, after 200 000 iterations 99.963%, and then 300 000 iterations 99.961%.

It looks like we're converging to the right answer. Note that we haven't proven this conclusion, we just keep narrowing the confidence interval as we increase the number of iterations. A nonrigorous but usually reasonable way of *testing* whether or not you've run enough iterations of a Monte Carlo simulation is to

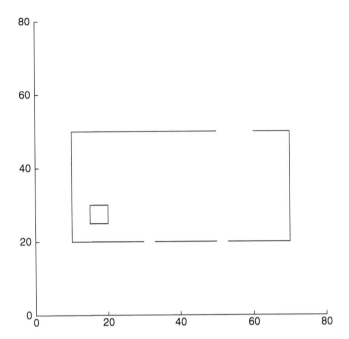

Figure 7.4 Geometry of room for Monte Carlo simulation.

just keep increasing the number of iterations by large factors and watch for the answer to "settle down."

Incidentally, this example can be solved easily analytically just by "thinking in reverse." That is, ask the question "what is the probability that no two single-digit integers in a group of 10 single-digit integers are the same?" For the first integer you have a choice of 10 out of 10, i.e. pick any integer between 0 and 9. For the second integer you must choose one of the remaining 9 integers (out of 10). For the third you have to choose one of the remaining 8 integers out of 10. Writing this out

$$\left(\frac{10}{10}\right)\left(\frac{9}{10}\right)\left(\frac{8}{10}\right)\left(\frac{7}{10}\right)\left(\frac{6}{10}\right)\left(\frac{5}{10}\right)\left(\frac{4}{10}\right)\left(\frac{3}{10}\right)\left(\frac{2}{10}\right)\left(\frac{1}{10}\right) = .00036288 \quad (7.3)$$

The probability of getting at least one repeat is therefore $1 - .00036288 = .99963712 \sim 99.96\%$.

The second Monte Carlo Simulation is of a situation that can't be done better, or even at all, by calculations. Figure 7.4 shows a large rectangular box which represents a simple house floor plan. There are three openings representing open windows and an open door. The small square in the lower left represents a person.

The simulation will be to release mosquitos at the door and windows and then see how often a mosquito encounters the person. We need to create a *model* of how a mosquito flies: if the mosquito wanders out a door or window without encountering the person, we score this as "no encounter" and start another mosquito. If the mosquito touches the rectangle representing the person, we score this as "encounter" and start another mosquito.

If the mosquito reaches a wall, we back it away one step and take a new random step. This simulates the mosquito "bouncing off" the wall.

While simple *x* or *y* motions are good for explaining the concept of a random walk, a resulting two-dimensional walk doesn't look very realistic. Instead, we'll give the mosquito a random walk modeled by a normal distribution with the mean set to the direction of the previous step and a sigma of 20 degrees.

We could get very sophisticated and give the mosquito a sense of smell to direct its flight, using its distance from the person and direction to the person to modify the probability for its next step, but we'll opt for a simpler simulation and not do this.

What is the potential value of this type of simulation? By repeating it for, say, 1000 mosquitos, we can calculate the probability of a mosquito encountering the person. Then we could move the person and repeat the whole simulation. After a few of these moves and repeats, we would know the location in the room where the person is least likely to encounter a mosquito (and get bitten). We could do the same thing by leaving the person in place and moving the door and windows.

If our model resembles physical reality closely enough, we have a design tool for laying out rooms to minimize harassment from mosquitos without having to close all the doors and windows.

Figure 7.5a–d show a few examples of (the results of) this simulation. In the trajectories shown, "o" marks the spot where the mosquito entered the room, and "x" marks the spot where it was extracted.

A League of Our Own

This last example of Monte Carlo simulations exemplifies a class of problems that have many solutions, some better than others. We will introduce an important concept called the Figure of Merit (FOM).

In this type of problem, there is only a finite probability that we ever get to the best solution to our problem or even know if there is a unique best solution. We only care that we are constantly getting a better solution (and can measure that we are indeed getting better).

The Acme Ephemeral Company has an in-house baseball league. There are five teams. Each team has a coach, a full roster of 9 players, and a few extra

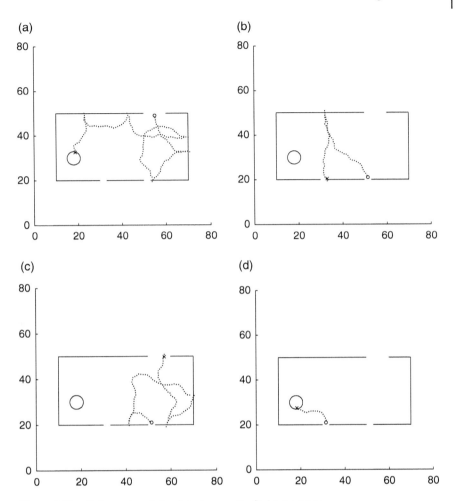

Figure 7.5(a–d) Examples of simulated mosquito flights in room.

players scattered here and there among the teams. On these teams the pitchers are unique (they can only be pitchers) and the catchers are unique (they can only be catchers), but the infield players are fungible (each infield player can play any infield position) and the outfield players are similarly fungible. The league has been running for a few years now; several players have dropped out and several new players have joined. The problem is, the games are getting stale – most of the players have been on the same team, with the same coach and with many of the same players for most, if not all, of the years that the league has been in existence.

We want to scramble the players so as to minimize this "staleness." That is, we want each player to be placed with as many other players and with a coach that they've *not* been together with before as possible.

The first thing we have to realize is that there is no perfect solution. Each team, including the coach, is made up of at least 10 people. No matter how we distribute people among 5 teams, there is no way to keep some people from being with people they've been with before. Acknowledging that we can make things better but never perfect, we'd still like to proceed.

While it's easy to define the perfect rearrangement, even though this is impossible to achieve, the definition of a "better" rearrangement that's short of perfect will require some discussion.

We create a "trial" league as follows:

1) Randomly assign each coach to a team, 1 coach per team.
2) Randomly assign each pitcher to a team, 1 pitcher to a team. If there are excess pitchers, randomly assign them to a team, 1 pitcher to a team, until we run out of pitchers or there are 2 pitchers on every team. Repeat this procedure until we've run out of pitchers.
3) Repeat the pitcher process for the catchers.
4) Randomly assign all infielders to teams, 4 infielders per team, then follow the excess player procedure as you did for the pitchers.
5) Randomly assign all outfielders to teams, 3 outfielders per team, then follow the excess player procedure as you did for the pitchers.

We now have a league that satisfies all the restrictions that are necessary to make all the teams functioning teams (they each have at least one coach, one pitcher, one catcher, 4 infielders, and 3 outfielders). Now we have to evaluate how well this league fits the rearranging goals.

We create a number called the Figure of Merit (FOM). The definition will not be unique – we'll suggest some variations later. For now, just take it as a number whose starting value is 0.

Take every possible pair of players (for this exercise, the coach is considered to be a player) on these trial teams and then search all the team rosters for the past 3 years looking for the same pair of players together on a team in a given year. Each time we find a match, add 1 to the FOM. When we're all done, we'll have a measure of how badly this trial league achieves the goal of a "total scramble." If we had not found any matches (this is impossible, as discussed above) the FOM would = 0 and we would have an exact solution.

Since this is the only trial league so far, this league and its FOM are the best we've ever created. Save the makeup of this league and its FOM.

Now, repeat the process. That is, randomly generate another league using the same procedure as above and then calculate the FOM for this league. If the new FOM is the same or higher than the first FOM, discard this latest league and start over again. If the new FOM is lower than the saved FOM, discard the saved FOM and league and replace it with the latest FOM and league.

As we repeat this process, we will be driving the saved FOM down. In other words, we're always working toward a league that more and more approaches the ideal scrambling goal.

This process never finishes. You never know if the next trial league won't be better (won't have a lower FOM) than the one you're saving. It's easy to see that there are many, many millions of possible leagues that satisfy the basic constraints, i.e. would be leagues with fully functional teams.

In principle, we could write a computer program that methodically steps through every possible combination of players and calculates the FOMs. This program would take a long time to run, but when it's done we know that we have found the league with the lowest possible FOM. It's easy to see that, while the simulation process is frustrating and doesn't always give us the best results, the direct calculation process can be too slow to be practical.

Discussion

FOMs are typically used in problems that are too involved to solve analytically. Often, every design decision to make one parameter a little better will make some other parameter a little worse. Only with a thoughtfully constructed FOM can the computer automatically make the best design tradeoffs.

In many cases the FOM is defined as a measure of "goodness" rather than a measure of "badness" as we did above. In this latter case, you work to maximize the FOM rather than minimize it. If you have savings and investments you have already been working this kind of problem. Your FOM is the dollar value of your total savings and investments minus your debts. You regularly make decisions (buy this, sell that, etc.) to maximize your FOM.

When your FOM is the dollar value of your total savings and investments, the FOM is uniquely defined. However, in more complex problems, it is always possible to put weighting factors on the different numbers that sum to create the FOM definition. In the baseball league problem, for example, we might decide that it's most important to separate pitchers and catchers that have been together in the past. To do this, we change the FOM definition so that we subtract, say, 2 rather than 1 to the FOM when we find a prior pitcher–catcher match. If it's really important, we could subtract 3, or 4, or whatever. Typically, there is a table of the relative importance of the different contributing factors and the weights assigned to each of them which can be "played with" to see just what the computer program produces.

When we're dealing with a problem that has continuous rather than discrete variables, the situation is usually handled somewhat differently. Suppose, for example, we're designing a television. The marketing people tell us that customers want a large screen but they also want the set to be as light as possible and of course as inexpensive as possible. Clearly as the screen gets larger, the weight and price will go up. We would design an FOM that takes all of these factors into account and then sit down with the marketing folks to get their estimates on how much potential customers would pay for a larger screen, etc., and translate this information into FOM weighting factors. The engineering design team will turn the relationships between the size, cost, and weight into equations. There are many mathematical techniques for formally searching the set of equations for optimum values to push the FOM down (assuming that we defined the FOM so that smaller is better). Some of these techniques are very smart in terms of looking for the absolute smallest (called the "global" minimum) value of the FOM without getting stuck in regions of "local" minima. However, none of them can guarantee that they've found the best possible answer in a very complicated problem, and all of them require some oversight in choosing starting values of the variables.

Many of today's computer spreadsheet programs offer FOM calculations minimizing (or maximizing) capability; the user defines the FOM in terms of relationships between the values in different cells of the spreadsheet. The step size guesses are usually suggested by the computer but can be overridden by the user. The relative weighting factors are provided by the user in building the formula for the spreadsheet cell that defines the FOM.

Some of the mathematical FOM search techniques available today are conceptually based upon natural processes. For example, there's an "annealing" technique that's modeled on the movement of molecules of metal as a liquid metal cools to its final (solid) state; there's a "genetic" technique that's modeled on evolutionary random mutations; and so on.

As was suggested above, these processes have as much art as they do science. Even assuming that we've found the best possible computer algorithms for optimizing our FOM, the results depend on the weighting factors put into the FOM and the starting values of the variable. Therefore, to some extent the overall value of the results is determined by the person(s) setting up the calculation.

Notes

Creating Monte Carlo simulations are computer programming exercises. The basic ideas were shown in Chapter 4. While it is easy to create many interesting and challenging problems, solutions would be discussions of programming alternatives without much probability discussion.

For example, we could simulate gas molecules ricocheting about in a box in a low pressure region (no particle–particle collisions) and a higher pressure region (particle–particle collisions). We could extend this to looking at kinetic energy distributions, pressure at the walls of the box, entropy, etc. We would be looking at particle collision mechanics, energy and momentum exchange, etc. This is a statistical mechanics problem with almost no fundamental probability issues in the simulation program.

The bottom line here is that there are no problems provided for this chapter.

8

Some Gambling Games in Detail

The Basic Coin Flip Game

The basic coin flip game is an excellent vehicle for examining probability calculations, their results, and implications. Remember, the game is simple: tails give you a dollar to your opponent, heads your opponent gives you a dollar. Assume a fair coin, that is, $P_{head} = P_{tail} = \frac{1}{2}$. The expected value (EV) of return after a game is simply

$$E = (winning\ amount)(P_{winning}) + (losing\ amount)(P_{losing})$$
$$= (\$1)(.5) + (-\$1)(.5) = 0 \qquad (8.1)$$

The EV after any number, n, of flips must similarly be zero. This is what we are calling a *fair game*.

Figure 8.1 shows (the upper and lower bounds of) the 95% confidence interval of these expected winnings versus the number of coin flips.

These numbers are the one-dimensional random walk "couched in coin's clothing." As in the random walk example, the standard deviation, $\sigma = \sqrt{n}$, n being the number of coin flips. The expected earnings are 0, but the confidence interval gets wider and wider with increased number of coin flips. At $n = 50$, for example, you are 95% confident of going home with somewhere between 14 dollars won and 14 dollars lost. At $n = 100$, you can be 95% confident that you will go home with somewhere between 20 dollars won and 20 dollars lost.

Suppose you start out lucky and your first 2 flips yield heads. This is not improbable (1/4 probability). You now have 2 dollars of winnings in your pocket.

Going forward, the coin has no memory of your earnings (or equivalently, of its own or of another coin's history). This means that you can expect earnings of 0 going forward, despite the fact that you have just won 2 dollars. Putting this a little differently, your total expected earnings for the evening is no longer 0, but is 2 dollars! This can be generalized to the statement that your expected

Probably Not: Future Prediction Using Probability and Statistical Inference, Second Edition. Lawrence N. Dworsky.
© 2019 John Wiley & Sons, Inc. Published 2019 by John Wiley & Sons, Inc.
Companion website: www.wiley.com/go/probablynot2e

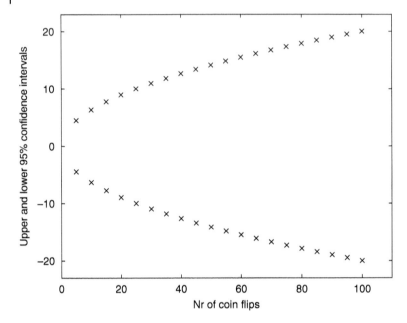

Figure 8.1 95% Confidence intervals versus number of coin flips, simple coin flip game.

earnings for the evening is exactly your earnings at that point in the evening (we are considering losses to be negative earnings). When you quit for the evening, your expected earnings is then your actual earnings.

To help make this point clear, consider Table 8.1. Table 8.1 shows the above situation for a 4 coin flip evening. There are 16 possible scenarios. Of these 16, only 4 have Heads for the first two games. This agrees with the simple calculation that getting 2 heads from 2 coin flips has a 1/4 probability. These 4 games are shown on top of the table, the remaining 12 games follow. Once the first two games are won, these latter 12 games are irrelevant and we need only consider the first 4 games. Now, look at the last 2 flips of these first 4 games. They show winnings of +2, 0, 0, −2, leading to an EV of 0. The evening's game winning possibilities are then +4, +2, +2, 0, with an EV of +2.

Looking at this situation by eliminating lines from the table leads to a proper interpretation of the situation. Winning the first 2 (or any number of) games does not change the future possibilities. Again, coins have no memory nor knowledge of each other's history. What winning the first 2 games does is to eliminate from the list all of the possible games in which you did *not* win the first two games.

Returning to 95% confidence interval, assuming you just won 5 dollars. It's the same curve as before except that 5 dollars has been added to every point. *n* now means the number of coin flips ahead of you, not counting the 5 flips that have already won you $5. Figure 8.2 shows this.

Table 8.1 Coin flip game, expected winnings when first two flips are won.

1	2	3	4
H	H	H	H
H	H	H	T
H	H	T	H
H	H	T	T
...	...	...	...
H	T	H	H
H	T	H	T
H	T	T	H
H	T	T	T
T	H	H	H
T	H	H	T
T	H	T	H
T	H	T	T
T	T	H	H
T	T	H	T
T	T	T	H
T	T	T	T

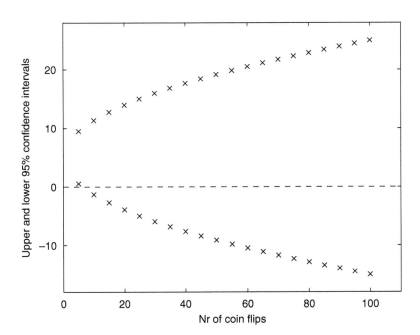

Figure 8.2 Repeat of Figure 8.1, starting with $5 winnings.

For $n = 5$, we have the case that the 95% confidence interval is from (approximately) 0 to +10. In other words, if you have already flipped the coin 5 times and have won 5 dollars, and then plan to flip the coin 5 times more, you can be very confident (in this case absolutely confident) that you will go home with somewhere between 0 and 10 dollars profit. This is a quantified statement of common sense: If you have $5 in your pocket that you didn't start out with and you are willing to flip the coin only five times more then there is no way that you can, in total, lose money.

Looking a bit closer, $n = 5$ is a small enough number for there to be inaccuracies in approximating a binomial distribution with a normal distribution. For $n = 5$, we do not have a 95% or 98% or any such confidence interval that you can't lose money. It's a certainty that you can't lose money.

Let's pretend we're gambling casino owners looking to introduce a new product: a 10 Flip Coin Game. In a given week, we hope to bring in 3000 people flipping coins. The probability of someone getting 10 heads in a row is 1/1024. Using the binomial PDF formula for $n = 3000$ and $p = 1/1024$, we find that the probability of at least one person getting 10 heads in a row is 95%. This is – of course – what happens when you give a very unlikely event enough opportunities to occur.

The mean ($\mu = np$ for a binomial distribution) of this distribution is for 3 people to flip 10 heads in a row, with a sigma of 1.7. (The distribution is an excellent approximation of the Poisson distribution.) We can be very sure that we won't see many people out of the 3000 doing this (this probability is about .0007), so why not offer $200 back on a $1 bet to anyone willing to try? With so many people passing through every day, we can be very sure of a reliable, comfortable stream of income.

The above example typifies the Las Vegas casino strategy: set the odds (and watch the sigmas) so that you're very sure that you are both returning terrific winnings to a few people while maintaining a healthy profit margin at the same time. The fact that it's inevitable for a few people to "win big" every night seems to make a lot of people think that it's worth taking the risk to lay down their money. There's even a strong temptation to carefully watch last night's big winners and to try and understand "their system." The casinos often have loud bells and flashing lights that go off when somebody wins one of these games, just to put the idea into everybody's head that they could be the next big winner.

Now let's look in a different direction. Suppose we have a weighted coin that has a probability of 60% of landing heads-up. Playing the same game as above, the expected winnings are just the EV per game times n, the number of games:

$$Winnings = n((.6)(\$1) + (.4)(-\$1)) = n(\$.20) \tag{8.2}$$

This looks good; the more you play, the more you win. Figure 8.3 is a graph of the upper and lower 95% confidence intervals.

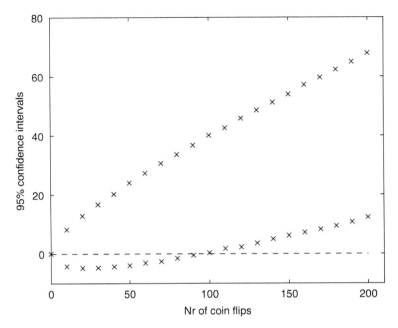

Figure 8.3 95% Confidence intervals versus number of coin flips using a weighted coin.

This game is a gambler's dream. Except for the cases of flipping less than about 100 times, you can be very confident that you won't lose money. And, the more times you flip this coin, the better off you are.

Now let's play this same weighted coin game with one additional constraint: you don't have infinite resources. What if you showed up at the casino with very little money in your pocket (a small "bankroll")? You know that if you can stay in the game long enough, the statistics make winning just about inevitable. But, what is your risk of getting "wiped out" before you have won enough to not have to worry?

Figure 8.4 shows the probabilities in this situation, assuming that you start out with $1 in your pocket. n is the number of times you flip the coin before you "call it a night." For $n = 1$ there's a .4 probability of getting wiped out. This makes sense – if the probability of winning with this coin is .6, then the probability of losing 1 flip has to be $1 - .6 = .4$. For $n = 2$ there's also a .4 probability of getting wiped out. This is because with 2 coin flips there are 4 possibilities, 2 of which wipe you out after the first flip – and the probability of a first flip wiping you out is .4.

As n increases, the probability of getting wiped out also increases. The pattern of the probability for any odd number being the same as the probability for the next even number continues. The overall probability of getting wiped out levels off at about .67.

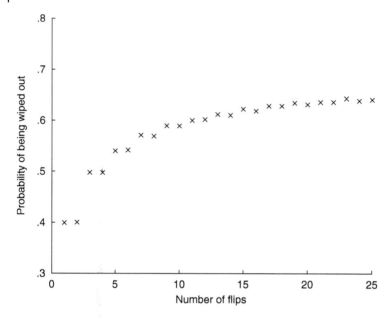

Figure 8.4 Probability of being wiped out with a weighted coin and finite starting resources.

The graph of the probability of getting wiped out versus the number of coin flips looks the same regardless of the starting bankroll, but the leveling out, or *asymptotic* value of the probability for a large number of flips drops rapidly as the initial bankroll is increased. Table 8.2 shows this plateau value for different initial bankrolls.

The message thus far is pretty clear: if the odds are in your favor, you can expect to win in the long run. However, you need to start with "deep enough pockets" to survive short-term uncertainties.

From the gambling casino's point of view, *they* are the people with the odds in their favor, and they have very, very big bankrolls (aka deep pockets) to carry them through statistical fluctuations. This is a good business to be in.

The "Ultimate Winning Strategy"

The next game is a game that you seemingly can't lose, but for some reason you just can't seem to win.

Using your fair coin, bet $1 on heads. If you win, you've won $1. Call this a success and say that the game is over. If you lose, bet $2 on heads and try again. If you win, you've covered your earlier loss of $1 and have won $1. Again, call

Table 8.2 Weighted coin flip game, wipe out probability versus starting bankroll.

Bankroll ($)	Wipeout probability
1	.67
2	.44
3	.29
4	.19
5	.12
...	
10	.01
25	<1 in a million

this a success and say that the game is over. If you lose this second flip, bet $4 on heads and try again. If you win, you've covered the $1 loss of the first flip, the $2 loss of the second flip, and won $1. Once again, call this a success and say that the game is over. Keep this procedure running until you've won your $1.

This scheme is unbeatable! How long can it take before you get a head when flipping a fair coin? In a short time you should be able to win $1. Then start the whole game over again, looking only to win (another) $1. Keep this up until you're the richest person on the earth.

Now let's do the real analysis; we add in the constraint that you can only enter this game with finite resources in your pocket. Let's say that you walk into a game with $1. If you lose, you're broke and have to go home. You can't flip the coin a second time. In order to be able to flip the coin a second time, you need to have $1 for the first flip and $2 for the second flip, or $3. If you lose twice, you need an additional $4 for the third flip, i.e. you need to have $7. For a fourth flip (if necessary) you need to have $15, etc.

The need to walk into the game with all this money to cover intermediate losses is alleviated a bit if you win a few games at the outset – if you start out with $1 and win 2 games immediately, you now have $3 in your pocket and can withstand a game that requires 2 flips before winning, etc.

Putting all of this together, we get an interesting set of curves. Figure 8.5 shows the probability of walking away with up to $10 if you walked into the casino with $1 in your pocket. Another way of putting this is the probability of winning up to 10 of these games in a row if you walked into the casino with $1 in your pocket.

The first point on the left of Figure 8.5 is the probability of walking away with $1. This is a no-brainer. You walk in to the casino, turn around, and walk out. There's no way that you can lose your money, so your probability of success is exactly 1.

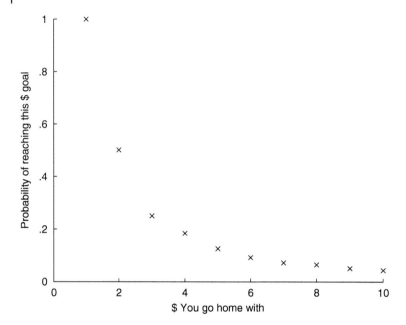

Figure 8.5 "Ultimate Winning Strategy" probabilities of winning starting with $1.

The second point is the probability of walking away with $2. You flip a coin once. If you lose, you're broke and you must go home. If you win, you have $2 and you choose to go home. Your probability of success here is exactly .5.

The remaining points are the probabilities of walking away with $3 or more. As can be seen, these probabilities decrease significantly as the amount of money you hope to win increases.

Now we'll repeat the last graph, but add the probability of winning $10 if you start out with $4 or 7$ instead of $1 (Figure 8.6).

The first case of the new sets of points mimics the same case for the original set of points (Figure 8.5). If you want to go as you came, turn around and leave (100% certainty). You have more resilience starting out with $4 than you did starting out with $1 and you can expect better "luck." As the winnings numbers get high, however, even though you are better off starting with $4 than you were starting with $1, the probabilities still get very small. This same argument extends to starting out with $7.

Looking at Figure 8.6, you can see that the probability of going home with $8 if you came in with $7 is very high. This is because you had a lot of money to back a very modest goal.

The system cannot be finessed by clever strategies. For example, suppose I start out with $4 and have a goal of coming home with $10. I can play the game as described above, I can play two games with a goal of $5 each, etc. It doesn't matter.

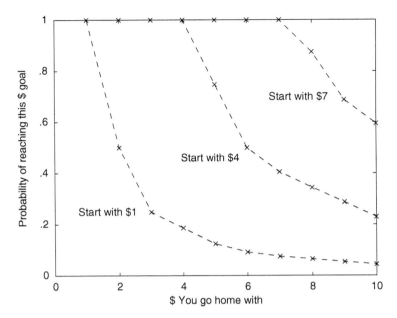

Figure 8.6 "Ultimate Winning Strategy" probabilities of winning starting with $1, $4, or $7.

The messages of this chapter thus far are easy to summarize:

1) In a fair game (EV = 0), you're playing with equal odds of winning and losing – there is no system or strategy that can help you.
2) In a game with an EV greater than 0 you will ultimately win, but make sure you are starting with enough in your pocket to cover early variations.
3) If you are lucky and win the first few games you play, then your prospects for the rest of the evening look better than they did when you first walked in. Similarly, if you lose the first few games you play …
4) The EV is the most likely result of any game but the confidence interval tells you just how sure you are of realizing this value.

Parimutuel Betting

In the gambling games discussed so far, the odds were fixed before you started playing. In the case of a roulette wheel we can calculate the odds by inspection. In the case of a slot machine we (usually) don't know the odds exactly, but you can be sure that the machine owner knows them.

However, *Parimutuel Betting* is based on an entirely different approach to helping you part with your money. At a horse race track, for example, the odds of a given horse winning are estimated beforehand. The actual odds are calculated and posted once all the bets are placed.

Consider the example shown in Table 8.3. In this table we show a horse race with eight horses entered. The second column shows the amount of money bet on each horse. The sum of all of these bets is $110 000. Assume that the race track owners want $10 000 income from this race to cover their costs and profits. In practice, the track owners' share is probably determined by a combination of a fixed amount plus some percentage of the bets. It isn't necessary to understand how the track owners came to their income number, we just need to know the number.

After deducting their $10 000 from the $110 000 betting revenue, the track owners have $100 000 left to distribute. This is very different from, say, a roulette wheel game where since the odds are fixed before the bets are placed the gambling casino owners only know how much they will distribute (and how much they will keep) each night as a statistical generality.

Column 3 of Table 8.3 shows how much money is paid to each gambler, per dollar of bet if her horse wins. The number in column 2 times the number in column 3 (for example $60 000 \times 1.67$ for horse number 4) always equals $100 000. This is just another way of repeating that $100 000 is always paid out to the betters.

The number in column 3 is also identically the odds on this horse as a ratio to 1. For example, the odds on horse number 1 was 50 : 1. The odds on horse number 4 is 1.67 : 1; this is more usually quoted as the equivalent ratio of two integers, 5 : 3.

Now, "odds" is defined as another way of quoting a probability with both representations being valid (sort of like quoting a cost in dollars or euros). In this situation, however, the meaning of discussing "odds" or "probability" of something after it's happened is a little hazy. Probability is the likelihood of an event chosen out of the allowed list of random variables with a given PDF that will occur *in the future*. If I flip 10 coins and get 10 heads, does it really mean

Table 8.3 Parimutuel betting example.

Horse number	Bets ($)	Return per $ bet ($)
1	2000	50.00
2	6000	16.67
3	8000	12.50
4	60 000	1.67
5	22 000	4.55
6	5000	20.00
7	6500	15.38
8	500	200.00

anything to discuss the probability of it having happened? It has already happened. When we believe that probabilities don't change in time, as in the probability of a coin flip, then we can use history to teach us what the probability is going forward. We're often a little sloppy with the language and will say something like "The probability of you having flipped 10 coins and gotten 10 heads was only 1/1024." What we really mean is "The probability of you flipping those 10 coins again and getting 10 heads is 1/1024." Something that has already happened doesn't have a probability, if it already happened it's a certainty that it happened. (A funny sentence structure that nonetheless is accurate.)

Before a race there are estimated odds published. These are best guess probabilities. After the race, the "odds" are really just a statement of how much money you would have won per dollar bet if your horse had come in.

Looking at the table again, we see that horse number 4 wouldn't pay very much if it won. This is because so many people thought that horse number 4 would win (or equivalently a few people were so sure it would win that they bet a lot of money on it) that the $100000 winnings had to be spread very thinly. Horse number 4 was the "favorite" in this race.

On the other hand, almost nobody bet on horse number 8. This horse was a "long shot" and if it won, would have made a few betters very happy.

From the perspective of the race track owners, this is a terrific system. So long as enough people show up to cover the owners' cut, the purse for the horse owner's plus some reasonable amounts to pay back to the betters, the track owners know exactly how much money they'll make every day. They never have to worry about a few slot machines "deciding" to pay off an awful lot of money one day.

Related in concept to parimutuel betting is the concept of *value pricing*. Value pricing is used by airlines and some hotels to adjust prices for their products as a function of the immediate market demand for the product. This is a significant extension of the idea that a resort hotel will have different "in-season" and "out-of-season" rates.

An airline will sell tickets at a fairly low price a long time before the date of the flight(s). This is because they haven't really established the need for these seats yet and also because this is one way of assuring your business and in turn giving the airline a better chance to project revenues and possibly even tweak flight schedules and the size of the plane.

On the other hand, an airline will charge very high prices for its tickets from about 1 to 3 days before a flight. This is when the business traveler who developed a last-minute need for a trip is buying tickets; this trip can't be postponed. In other words, the demand is very high for these seats and the customers can't walk away from a high price.

It's been said that no two people on a flight paid the same amount for their ticket. This is probably not literally true, but there certainly are many different prices paid for the tickets on a given flight.

Value pricing is related to parimutuel calculations in that the adjustments are being made in response to incoming data in real time. However, the computer algorithms that adjust the prices are not reflecting simply mathematics such as in the horse race example, but are programmed with formulas that reflect the airlines' marketing experts view of what people will pay for a ticket in different circumstances (and also in reaction to what the competition is charging). In other words, while there really isn't a parimutuel calculator setting the prices, it certainly seems that way when you're trying to shop for tickets.

The Gantt Chart and a Hint of *Another Approach*

Another example of anticipating (or not anticipating) variability in business planning is the Gantt Chart. The Gantt Chart is a chart with time as the horizontal axis and various activities needed to achieve some business goal (e.g. bringing a new product to market or getting a new factory up and running) on the vertical axis. A very simple Gantt chart is shown in Figure 8.7.

For complicated projects such as getting a high-tech factory up and running, the Gantt chart is typically an assemblage of dozens of sub-charts and the total package can be a very formidable document.

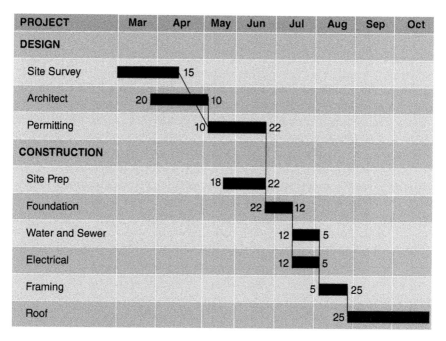

Figure 8.7 Sample Gantt chart.

The Gantt chart is an excellent management tool. Across the top is a date-line. Down the left side is a list of tasks. The bars show the planned start and completion dates for each task. The numbers on either side of the bars tell the actual dates of the planned beginning and end of each task. The lines connecting the ends of tasks to the beginnings of other tasks are dependency lines – that is, indicators of what activities can't start until previous activities have finished.

In this very simple example, the site survey, the architect, and the site prep could start at any desired date. Permitting, on the other hand, cannot begin until the surveyor and the architect have completed their tasks.

Realizing that in business, time is indeed money, Gantt charts can be very useful in predicting when a project will be completed and when specific resources (architects, site excavation crews, etc.) will be needed.

This chart was drawn as a plan before the project was started. Do we know, for example, that the survey (the top item on the chart) will actually be completed on April 16th? What if it rains suddenly and the schedule slips? What if the surveyors have a car accident on their way to work one day? What if the surveyors' previous job runs late and the start of this job gets pushed back?

All we can do is predict (from experience) that the survey activity will take 46 days with a standard deviation of, say, 4 days, and that the survey activity will begin on March 1 with a standard deviation of, say, 2 days. In this sense, Gantt chart planning can be looked at as related to gambling. It's a scenario with a set of steps; means and sigmas are assigned to each step, and we predict our probability of winning. "Winning" here doesn't meaning winning a certain amount of money, it means completing the job on (or before) a certain date.

Putting a sigma on every activity doesn't change the predicted completion date. However, it does give us a composite sigma for how well we know this date. This in turn lets us calculate our confidence interval for the date of completion of the job. The answer may be painful to senior management who will stamp their feet, throw coffee cups, and threaten to "bring in someone else who can tell us that this job will be done on time." In this latter situation, realize that without more information, all "someone else" can do is tell you whatever she feels like telling you, she has not changed anything.

Since the final distribution is in all likelihood normal, if the input information is good then there's a 50% probability that the job will be done early rather than late. There's a 25% probability that a manager will be able to bring in two projects in a row that are early. This manager will look better than two managers who each brought in one project early and one project late, and will look much better than a manager who brought in two projects in a row late (note of course that all 4 cases were equally likely). When a promotion opportunity opens up, should this first manager be chosen because he/she has a "track record"?

Shouldn't we realize that elements of chance can often weigh seriously into the results of what at first glance appear to planned, deterministic events, and not be too eager to assign blame or reward for some of the variations?

Mathematically, there's no difference between looking at what's ahead when you've won your first two flips in a coin flip and looking ahead when you've brought in your first two projects ahead of schedule. Once you've assigned the probabilities to each job task (or each coin flip) and know how much money you're starting out with (or how many days of slack you have in your starting date), the final distribution function and the confidence intervals are determined.

This is where we take our first look behind a door that hasn't been opened yet. What if we stop thinking of the probability of an event as a fixed number and start thinking of it as our best guess as to what the number is? Is there a mechanism whereby real world experience can be used to upgrade our initial assumptions? Jumping back to simple coin flips, suppose you flipped a coin ten times in a row and saw ten heads. The fair-coin assumption says that the probability of this happening is approximately one in a thousand. This is unlikely, but certainly not impossible.

However, suppose that you did just flip ten heads and then had to bet on the results of the eleventh flip? If you were absolutely certain that you were flipping a fair coin, the probability of an eleventh head is one-half. Is that how you'd bet, or at this point would you start getting suspicious about the true nature of the coin and bet on an eleventh head?

Returning to our management promotion choice, we know that assuming that the results of all the tasks in all of the projects are random leads us to the conclusion that all four managers are just victims of random circumstances and we can choose any one of them randomly for the promotion. But, just as in the coin toss example, shouldn't there be some way of quantifying the fact that maybe all the circumstances weren't random (just like maybe the coin isn't really a fair coin) and maybe the manager who brought in both projects early is the best choice?

Yes, there is such an approach. This approach to probability is the subject of Chapter 11. It doesn't invalidate everything presented so far. Rather, we will see that allowing experience to update probabilities is an important different way of looking at things.

Problems

8.1 Let's try a new strategy for our basic coin flip gambling game. Since our EV of earnings for the evening goes up every time we win and goes down every time we lose, might we not be smart to raise our bet every time we win and reduce our bet every time we lose?

Start out with a lot of money in your pocket (i.e. don't consider getting wiped out just yet). Bet $1. If you win, make your next bet $2. If you lose, make your next bet $.50. Continue this strategy: if you win twice in a row, your third bet is $4, if you lose twice in a row your third bet is $.25, etc.

A What is your EV of winning versus the number of coin flips?

B What is your 95% confidence interval versus the number of coin flips?

C If you start out with only $1 in your pocket, what is the probability of getting wiped out versus the number of coin flips?

8.2 A car dealer donates a new car with a value of $35 000 to some charity event. The event's organizers decide to sell raffle tickets for the car. They will sell 2500 tickets at $10 each. Setting aside philanthropic intentions for the moment, discuss the gambling aspects of this situation: is it a good idea to buy a raffle ticket?

8.3 A Referring back to Problem 8.2, how many raffle tickets would you have to buy before the odds of winning the car are in your favor?

B If you could find 2500 identical raffles and buy 1 ticket at each of them, what is your probability of winning at least once?

C How many tickets would you have to buy at each of the 2500 raffles to have a 95% probability of winning at least one car?

8.4 In this problem we will look at a ticket purchaser's EV of return in a hypothetical state lottery. The lottery rules to be presented may or may not actually exist in a real lottery. The numbers here have been chosen to keep the calculations modest (i.e. make it practical to solve this problem using only a spreadsheet), real state lotteries are usually a lot bigger – the purchaser has more choices, the probability of winning is much lower, the number of tickets sold and the prizes are much bigger.

In our lottery, each ticket costs one dollar. The state keeps 30 cents from each ticket and puts the other 70 cents into a prize pool. The prize pool therefore grows directly in proportion to the number of tickets sold.

Each ticket holder, at the time of buying the ticket, chooses three different numbers between one and fifty.

A big bucket holds fifty balls, numbered consecutively from one to fifty.

At the lottery's conclusion, a lottery official draws three balls randomly from the bucket. If a ticket holder's three numbers match three of the five drawn ball numbers, the ticket holder is a winner. There may be no winners for a drawing; the state then sells more tickets for a period of time and has another drawing. If there is one winner, this winner gets the prize pool. If there is more than one winner, the prize pool is divided evenly among the winners.

You buy a lottery ticket. What is the EV of your return as a function of the number of tickets sold?

8.5 Note: Solving this problem requires some familiarity with linear algebra, at least insofar as setting up and then solving a set of linear equations. If you don't have this capability, you can still understand what the problem is asking and how the answers solve the problem; i.e. it's still worth reading through.

Parimutuel betting is often more complicated than what was presented in this chapter. In horse races, for example, there is a first horse to cross the finish line bet; a first or second horse to cross the finish line bet; and a first, second, or third horse to cross the finish line bet.

What makes the parimutuel calculation involved is that you do not just bet on, say, horse A to be first, horse A to be second, or horse A to be third. The calculations in this situation would be just three sets of the calculations presented in this chapter. You bet on

1 Horse A Win bet: horse A comes in first
2 Horse A Place bet: horse A comes in second or first
3 Horse A Show bet: horse A comes in third, second or first

The Win bet was covered in this chapter. This problem is about the Place bet. Consider the four horse race shown in the below table. The bets shown are the Place bets on each horse. Ignoring the money taken by the racetrack owner for costs and profits (i.e. return all money to the betters), show the amount each better will receive if his horse places.

Horse	Total bets
A	$1000
B	$2000
C	$3000
D	$4000

9

Scheduling and Waiting

Introduction

Many of our daily activities involve waiting. We wait for a bus, we wait to see our doctor in her office, we wait to meet a friend for lunch, etc. Wouldn't it be nice to have your bus pull up just a minute after you reach the bus stop, or have your doctor take you in for your examination just a minute after you reached her office? On the bus company's side, there is a tradeoff between minimizing the number of buses (operating cost), yet not overloading buses and not having people wait too long in possibly inclement weather. On the doctor's side there are clear advantages to getting to see as many patients as possible in a day while being able to go home for dinner at a regular time (yet not having patients waiting long times for treatment). When you meet a friend for lunch, the exact time of the meeting usually isn't critical, but if one of you arrives 15 minutes earlier and the other 15 minutes later than the agreed-upon time, one of you will have to wait and ponder the value of timeliness.

Let's start with the doctor's office and see what we can learn, next we'll look at meeting friends for lunch, and then at bus schedules.

Scheduling Appointments in the Doctor's Office

Assume that the working day is from 9 to 5, 8 hours long. Even though real people like to break for lunch, assume that our doctor works straight through. We're doing this because taking account of lunch breaks complicates the calculations and conclusions without adding anything to our understanding of the statistics of the situation. Also, we're going to ignore the possibility that someone calls at the last minute with a medical need that requires, if not immediate, at least prompt, attention. This can happen to a doctor and the best she can do

Probably Not: Future Prediction Using Probability and Statistical Inference,
Second Edition. Lawrence N. Dworsky.
© 2019 John Wiley & Sons, Inc. Published 2019 by John Wiley & Sons, Inc.
Companion website: www.wiley.com/go/probablynot2e

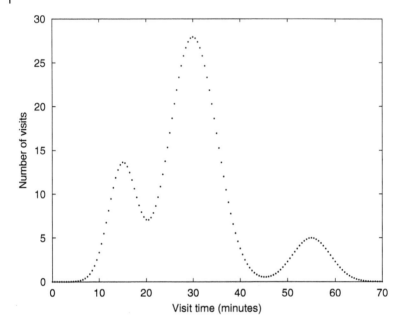

Figure 9.1 Hypothetical doctor's office number of visits versus visit time distribution.

is to juggle the remaining appointments and try to notify everybody of the changes before they commit to showing up at the office.

Finally, it is unreasonable for patient appointments to follow each other immediately; the doctor needs time to write up notes, etc. We'll absorb this into patient visit times and we will have appointments follow each other immediately.

At first, our doctor set up appointments every half-hour (30 minutes), 16 appointments for the day. She soon noticed that this led to very erratic timing issues in her office. And, she seemed to be finishing late most days.

The doctor decided to be scientific about this. She instructed her receptionist to record visit times. That is, record the time from when a patient walked in to the office from the reception room to the time they came out of the office.

When the receptionist had recorded about 1000 office visits, she plotted her data, as seen in Figure 9.1.

Figure 9.1 does not show a simple distribution function. There are three peaks, located at approximately 15, 30, and 55 minutes.[1] The EV is 30 minutes (to the nearest minute) and the standard deviation is 11 minutes (to the nearest minute), but it's not clear what we have learned by calculating these numbers.

---

1 This somewhat bizarre looking PDF is meant to capture a combination of very short visits (e.g. a follow up on blood pressure testing), a moderate length visit (e.g. a twisted ankle diagnosis), and a lengthy visit (e.g. a full physical examination).

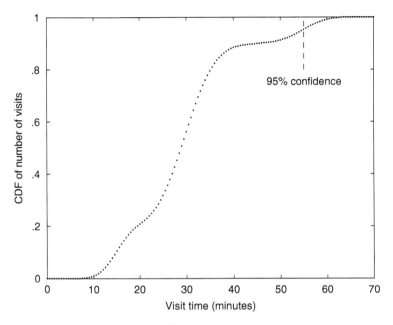

Figure 9.2 CDF of distribution in Figure 9.1.

Calculating the 95% confidence interval of being finished with a patient, assuming that our data is Gaussian, we should use 1.65σ as our estimate (this is a one-sided confidence calculation, not the more common two-sided calculation for which we'd use $\pm 2\sigma$), we get 48 minutes. This means the doctor could only schedule ten patients into her working day if she wanted to be 95% confident that she could go home at 5:00 every day. But, how accurate is this confidence factor calculation for such an un-Gaussian looking distribution?

Figure 9.2 shows the Cumulative Distribution Function (CDF) of the receptionist's data, with the 95% point shown. This point is at 55 minutes, not 48 minutes. This means that there is only approximately enough time for 9 patients in a day. This does not agree with the doctor's experience.

Is the above calculation correct? We're adding up nine worst-case situations to get to our conclusion. Let's look at it differently; add up the distribution functions so that we bring in some averaging for patients who take less than 55 minutes for their visit.

After adding together distribution functions, the result starts looking normal, and of course it looks more and more normal as we continue to add distribution functions.

Adding 13 of the original distribution functions, we get an EV of 384 and a sigma of 441 minutes. Since this is a normal distribution, we may calculate the one-sided 95% confidence directly, getting 451 minutes, about seven and a half hours. For 14 distribution functions, we get eight hours and six minutes.

At this point the doctor needs to make a decision. She can schedule 13 patients a day and meet her original criterion, or she can schedule 14 patients a day and be 95% confident that she will be finished seeing patients by about 5:05 and can go home.

Lunch with a Friend

In Chapter 3, we looked at the sum of two random numbers. But what about the difference between two random numbers? This is not something new, because of basic algebraic relationships. The subtraction $3 - 2$, for example, is equivalent to the addition $3 + (-2)$. That is, instead of subtracting two positive numbers we're adding a positive number and a negative number. The result is of course the same.

Why bring this up now? Because this is the way to attack problems of waiting. For example, suppose you and a friend are to meet at a restaurant at 1:00 for lunch. You tend to be early. Specifically, you'll show up anywhere from on time to 15 minutes early, with a uniformly distributed probability of just when you'll show up. Your friend, on the other hand, tends to be late. He'll show up any time from on time to 15 minutes late, again with a uniformly distributed probability of just when he'll show up. It isn't really critical at what time you start lunch, but there's that annoying gap between your arrival and his arrival during which you're wondering just how much your friendship is really worth.

Let's take one o'clock, the nominal meeting time, as our reference. Your arrival time is then a uniformly distributed random variable between 0 and −15 minutes and your friend's arrival time is a uniformly distributed random variable between 0 and 15 minutes. The probability distribution function (PDF) of the time gap between our arrivals is the difference between these times. We could work this out step by step or we could go back to Building A Bell (Chapter 3) and just look at the first example again. The resulting PDF is shown in Figure 9.3.

The minimum gap time, or waiting time, is 0, which happens when you both show up exactly on time. The maximum gap time is 30 minutes, which happens when he's 15 minutes late and you're 15 minutes early. The expected value, which is equal to the average because the curve is symmetric, is just 15 minutes. The standard deviation of each of our arrival PDFs is 6.1 minutes, and since our arrival times are uncorrelated, the standard deviation of the gap time is just

$$\sigma = \sqrt{6.1^2 + 6.1^2} = \sqrt{2(6.1)^2} = 6.1\sqrt{2} = 6.1(1.414) = 8.66(\text{minute}) \quad (9.1)$$

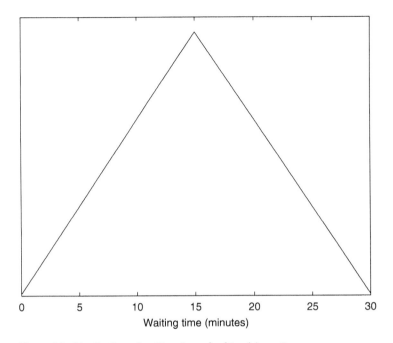

0   5   10   15   20   25   30

Waiting time (minutes)

Figure 9.3 Distribution of waiting times for friends' meeting.

Now let's change things bit. We'll make you both pretty good citizens in that on the average you both get to your appointments on time. However, neither of you are perfect in that you will arrival sometime within 7.5 minutes early and 7.5 minutes late. Again, we'll make the PDFs uniform. The PDF for your friend's arrival time now is the same as the PDF for your arrival time.

The time that one of you will have to wait for the other is just the difference between your arrival time and his arrival time. You have to subtract one PDF from the other. At the beginning of this section, we noted that subtracting a PDF is the same as adding the negative of the PDF. However, for a PDF that's symmetric about zero, the negative of the PDF is just the PDF again. In other words, there's no difference between adding and subtracting these two PDFs.

The result of combining these two PDFs is shown in Figure 9.3, just shift the horizontal axis so that the peak is at 0.

The question to answer is how long should the first arriver expect to wait for the second arriver? To answer this question you do not want the difference between the two arrival times but rather the absolute value of the difference between the two arrival times. That is, if he arrives 10 minutes early and you arrive 5 minutes late, the waiting time is 15 minutes. If you arrive 10 minutes early and he arrives 5 minutes late, the waiting time is still 15 minutes. There is no such thing as a negative waiting time in this sense.

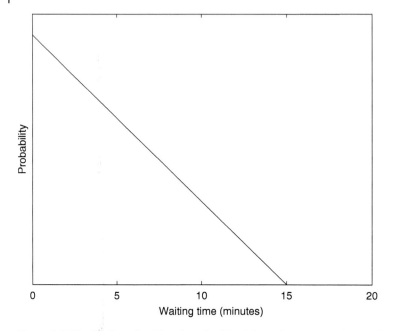

Figure 9.4 Distribution of waiting times for friends' meeting, symmetric arrival statistics.

Figure 9.4 shows the PDF for this new definition of waiting time. The minimum waiting time is 0 which occurs when we both show up at the same time, whatever that time might be. The maximum time is 15 which occurs when one of us is 7.5 minutes early and the other is 7.5 minutes late. It doesn't matter who does which. The expected value of the waiting time is 5 minutes and the standard deviation is 3.5 minutes.

We could redo this example, replacing the uniform distributions with normal distributions, or possibly with Poisson distributions. The results don't change in any surprising way. Fortunately, most good restaurants have bars to wait in.

Waiting for a Bus

Waiting for a bus is different than waiting to meet a friend for lunch. Waiting for a friend is symmetric in the sense that it doesn't matter if you or your friend get to the meeting place first. The situation is the same as the lunch meeting if you get to the bus stop before the bus and have to wait for your "meeting." On the other hand, if the bus gets to the bus stop (and leaves) before you arrive, there won't be any meeting – you'll have missed the bus. If there is only one bus that day and you have to get to your destination that day, then it is critical that you don't miss the bus. You have to look at the statistics of when the bus arrives as compared to the published schedule, and then you have to be honest with yourself about how well you meet a schedule. If we assume that both events are normally distributed with a mean of the time the bus schedule says the bus will arrive, then your target time for meeting the bus (with a 97.5% confidence factor) must be earlier than the bus schedule time by $2\sqrt{\sigma_{you}^2 + \sigma_{bus}^2}$. If you need a better confidence factor, then you must target getting to the bus station even earlier.

An entirely different problem arises when more than one bus comes past your stop each day. Assume that buses run hourly. In this case missing a bus can be reinterpreted as having 100% probability of catching the next bus. The interesting part of the problem is looking at what wait times will be and if there is any optimization (i.e. minimization) of your wait time to be had.

There are so many possible variants of this situation that it's impossible to consider all of them. We'll pick a few to examine.

Suppose that buses are supposed to come hourly, on the half-hour. There's a 1:30 bus, a 2:30 bus, etc. Due to varying traffic conditions, weather, road construction, etc., the buses almost never arrive when they're supposed to arrive. The 1:30 bus will arrive sometime between 1:00 and 2:00, the 2:30 bus will arrive sometime between 2:00 and 3:00, etc. You want to catch the 1:30 bus. Since you have no idea just when the bus will arrive, you get to the bus stop sometime between 1:00 and 1:59. What can you expect in terms of wait times and which bus will you actually catch?

Assuming all 3 PDFs (two buses and you) are uniformly distributed, there's a 50% probability that you catch the first bus and a 50% probability that you'll catch the second bus. Your shortest possible wait time is zero; this happens when the first bus pulls up immediately after you get to the bus stop. This can happen any time between 1:00 and 2:00. Your longest possible wait time is just under two hours – the first bus comes (and goes) at 1:00, you arrive at 1:01, the second bus comes at 2:59.

Figure 9.5 shows the PDF and the CDF for waiting times. This is a true continuous distribution, so it must be interpreted carefully. We cannot say, for example, that there is a probability of 1.0 of zero wait time. With a continuous PDF, we calculate the probability of an occurrence over a range of the random variable by calculating the area under the curve for this range. Usually this take integral calculus and the CDF provides the results of these calculations. For example, the probability that you'll have to wait between 0 and .5 hours is a little less about 40%. The probability that you'll have to wait between 0 and 1 hour is about 85%. The probability that you'll have to wait between .5 hours and 1 hour is about 35% because 85 – 50% = 35%. Also, while the PDF looks like the right-hand side of a normal curve, it is not (part of) a normal curve. One fact in particular supports this conclusion; this PDF goes to 0 at Wait Time = 2 hours, whereas a normal curve would slowly approach, but never reach, zero.

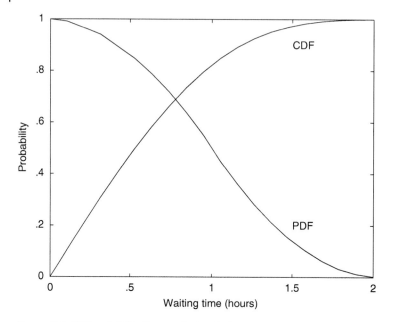

Figure 9.5 PDF and CDF of bus arrival wait times versus your arrival time, all arrivals distributed uniformly.

The PDF shows that it is much more probable that you'll have a short wait than a long wait.

In order to see if there is a best time to get to the bus stop, we'll change your arrival time at the bus stop from a random variable to a fixed time. Figure 9.6 shows the expected wait arrival times from 1:00 to 2:00 in 1/10-hour (6-minute) intervals. The expected wait time for arrival times of 1:00 and 2:00 are the same. This is because the situation at both of these times is identical – you arrived at the beginning of the hour with a bus definitely coming sometime during that hour. The worst time to come is at 1:30 because this time has the longest expected value of waiting time. This is an interesting result – the worst time to show up is the time on the bus schedule. All in all, it looks like you want to come just before the hour – just before 1:00 if you want to be sure to catch the first bus, just before 2:00 if you're OK with the second bus. Note that the worst case EV is about 40 minutes while the best case is 30 minutes; there isn't much optimization to be done here.

The discussions in this chapter about meeting a bus and waiting in a doctor's office point out that improvement in scheduling and waiting times can be had if there is good information about the variabilities involved. The discussion about meeting a friend is just an exercise in calculating means and standard deviations; the best situation is for both parties to show up as close to on time

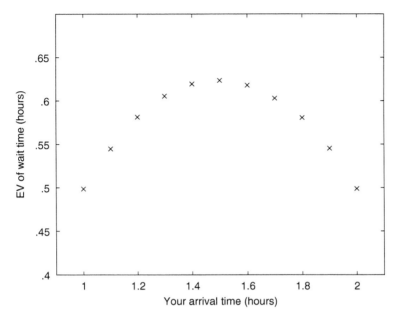

Figure 9.6 PDF of bus arrival wait times, your arrival time fixed.

as possible. However, in terms of the bus and the doctor's office, use of the PDFs involved show how some statistical information about the upcoming day's events can make the day better.

Problems

9.1 Repeat the Doctor's Office waiting time simulation using a uniform distribution between 15 minutes and 1 hour as the visit time PDF.

A few useful facts: The standard deviation of a uniform distribution is $\sigma = (.289)width$. For the sum of n uniform distributions, $\mu = \sum_{i=1}^{n} u_i$ and the standard deviation is $\sigma = \sqrt{\sum_{i=1}^{n} \sigma_i^2}$.

9.2 Repeat Problem 9.1, but assume the visit time PDF is a normal distribution with $\mu = .5$, $\sigma = .17$

9.3 This is the doctor visit scenario again, but from the perspective of the patients. Assume that Problem 9.1 (and its solution) reflect reality. What's the worst case (longest) case wait a patient can expect if he shows up on time?

9.4 Your company has a noon meeting every week.

   A If everybody attending the meeting shows up at a time distributed uniformly with a mean of noon and a width of 30 minutes, what is the latest you'll have to start the meeting based on how many people are attending the meeting?

   B What if arrivals are normally distributed about noon with a sigma of 10 minutes? Use a 95% confidence factor to determine the starting time.

9.5 The bus passes your house regularly every hour, but you don't know its specific schedule. What time should you arrive at the bus station to minimize the expected value of your wait time?

9.6 A bus that you want to catch comes once a day. Its nominal arrival time is noon. However, its arrival is a normal distribution with a mean of noon and a standard deviation of 10 minutes.

   A If I arrive at noon, what is the probability of my missing the bus?

   B If I want to be 95% confident of not missing the bus, what time must I arrive?

9.7 Buses on your route run nominally every hour and arrive at your stop (nominally) on the hour.

   A Each bus' arrival time at your stop has a normal distribution with each mean being the scheduled arrival time and all sigmas being 20 minutes. What is the probability that the 2:00 bus will arrive ahead of the 1:00 bus today?

   B If these buses are scheduled to run every hour on a 24-hour basis, what is the probability of the above happening once a day, more than once a day?

10

Combined and Conditional Probabilities

Introduction

All of the probabilities calculated up to this point have been, in a sense, *simple* probabilities. This doesn't mean that they were necessarily easy to calculate or understand, but rather that they all share some fundamental properties which will be explained below. In this chapter, we will talk about some not-so-simple probabilities, most of which won't be that hard to calculate and understand. We need to introduce some notation that can capture the ideas that we'll be presenting in a concise and precise manner.

Functional Notation (Again)

In Chapter 2, we introduced the idea of a function and then went on to Probability Distribution Functions. The common notation for a function is to use a letter followed by a variable, or variables, in parenthesis. Since we're dealing with probabilities, we'll stick with convention and use the letter P. If we're talking about rolling a pair of dice, we would represent the probability of a random variable, x, with the notation $P(x)$.

Then, the probability of the roll yielding 12 is

$$P(12) = \frac{1}{36} \tag{10.1}$$

This notation is a little bit confusing in that if we wanted to multiply a variable P by 12, we could write it the same way. The result of course most likely wouldn't be 1/36, depending on the value of the variable, P. How do you know whether we're talking about a function or a multiplication? From context – if there is a variable named P, then we wouldn't be naming a function P, and vice versa.

Probably Not: Future Prediction Using Probability and Statistical Inference,
Second Edition. Lawrence N. Dworsky.
© 2019 John Wiley & Sons, Inc. Published 2019 by John Wiley & Sons, Inc.
Companion website: www.wiley.com/go/probablynot2e

In this *functional notation*, we could make the above somewhat clearer by writing

$$P(x=12)=\frac{1}{36} \tag{10.2}$$

but this format is the exception rather than the rule.

The power of this notation is the ease with which we can express more complex expressions. For example, the probability of x being 10 or more is

$$P(x\geq 10)=\frac{3}{36}+\frac{2}{36}+\frac{1}{36}=\frac{6}{36}=\frac{1}{6}$$

With this notation as a working tool, we are now ready to start talking about *combined probabilities*. The first combined probability uses the logical operation *and*. Let x and y be the variables that represent the values of rolling each of the two dice. Then the probability that the red die will be greater than 4 and the green die will equal 2 is written as

$$P(x>4 \ and \ y=2) \tag{10.3}$$

As in all of our examples thus far, the two events ($x > 4$, $y = 2$) are independent. When events are independent, the probability of both of them happening (the *and* case) is just the product of the probabilities of the two occurrences:

$$P(x>4 \ and \ y=2)=P(X>4)P(y=2)=\left(\frac{2}{6}\right)\left(\frac{1}{6}\right)=\left(\frac{1}{18}\right) \tag{10.4}$$

Another logical operator is *not*. This is a convenient way to express things in many cases and is mathematically quite simple. Since an event must either occur or not occur, but not both, then

$$P(E)=1-P(not \ E) \tag{10.5}$$

where E is an event such as a die having a value greater than 3.

The last logical operator to consider was already used, just above. This is the operator *or*. The definition of the occurrence of event A *or* event B occurring means that either event A will occur, or event B will occur, or they both will occur.[1] As an example, what is the probability of a dice roll, dice labeled x and y, yielding ($x = 4$ or $y = 6$)?

---

1 There is also the *exclusive*-or which is defined as either event A occurs or event B occurs, but *not* both. This is the less common usage of the term *or*, and you can expect it not to be this case unless you're specifically told otherwise.

When die $x = 4$, there are 6 possibilities for y. When $y = 6$, there are 6 possibilities for x. There is one case where $x = 4$ and $y = 6$ simultaneously. $P(x = 4$ or $y = 6)$ is therefore 11.

This may be expressed more formally as

$$P(A \text{ or } B) = P(A) + P(B) - P(A \text{ and } B) \tag{10.6}$$

The subtraction on the right above corrects for counting a case twice.

A variation on this problem is the probability of rolling two unlabeled dice and asking for P(4 or 6) on either of the die. In other words, what is the probability of rolling two unlabeled dice and seeing at least one 4 or one 6? The simplest way to solve this is to make a list of the 36 possible combinations of rolling two dice and count how many of these combinations have a 4, a 6, or both. The answer is 20.

There are some situations where event A and event B are *mutually exclusive*. This is a fancy way of saying that the two events simply cannot occur at the same time. For example, the probability that the sum of two dice is less than 4 (event A) or the sum of two dice is greater than 10. There are 3 ways to get event A, and 3 ways to get event B, but no ways to get them both at once. The simple answer in this situation is therefore

$$P(A \text{ or } B) = P(A) + P(B) = \frac{3}{36} + \frac{3}{36} = \frac{1}{6} \tag{10.7}$$

because $P(A$ and $B) = 0$.

Conditional Probability

Several chapters ago we talked about the probability of winning money in a coin flip game once you had already won some money. We didn't name it at the time, but this is a case of *conditional probability*. Conditional probability considers the probability that some event B will happen once some interrelated event A has already happened.

The notation for this is $P(B|A)=$. This is read The Probability of B Given A, which in itself is a shorthand for The Probability of B Given That A Has Already Happened. This is different than the situations where the coin or the die has no knowledge of what the other coins or dice are doing, or even what it did itself last time. Now we're going to look at situations where interactions and history change things. The first few examples will be simple, just to show how things work. Then we'll get into some examples of situations and decisions that could literally mean life or death to you some time in the future.

Suppose we have 3 coins to flip and we want to know the probability of getting 3 heads. Before we flip any of these coins, we can calculate this probability as 1/8. Now we flip the first coin and get a head. What's the probability that we can flip the other two coins and get two more heads?

The original sample space consists of 8 possibilities (Table 10.1); the first possibility is the only one that shows 3 heads. Now we flip the first coin and get a head. This immediately eliminates the last 4 possibilities from the list and leaves the first 4 possibilities as a *reduced sample space*. There is still only one way to get 3 heads in this example, but now it is one way out of 4 possibilities. The probability of getting 3 heads once we've gotten 1 head, $P(3 \text{ heads}|1 \text{ head})$ is therefore ¼.

Generalizing, let's let event A be getting a head on the first flip, and event B be getting 3 heads after all 3 flips. The probability of getting 3 heads once I've gotten one head is the probability of B given A, written as $P(B|A)$. This must be equal to the probability of both events happening (B and A) divided by the probability of getting A. That is,

$$P(B|A) = \frac{P(B \text{ and } A)}{P(A)} = \frac{\dfrac{1}{8}}{\dfrac{1}{2}} = \frac{1}{4} \tag{10.8}$$

The same equation is often written as

$$P(B \text{ and } A) = P(A)P(B|A) \tag{10.9}$$

This is known as the *multiplication rule*. If B doesn't depend at all on A, for example the probability of a coin flip yielding a head after another coin flip yielded a head, then $P(B|A)$ is simply $P(B)$, and the above formula becomes the familiar simple multiplication rule for *independent* (i.e. not conditional) events,

$$P(B \text{ and } A) = P(A)P(B) \tag{10.10}$$

Table 10.1 Three fair coin flipping possibilities.

H	H	H
H	H	T
H	H	H
H	H	T
T	T	H
T	T	T
T	T	H
T	T	T

Since this general multiplication rule doesn't know which event you're calling A and which you're calling B,

$$P(B \text{ and } A) = P(A)P(B|A) = P(A \text{ and } B) = P(B)P(B|A) \qquad (10.11)$$

Consider the rolling of two dice. Each die is independent of the other, so we might not expect the above conditional probability relationships to be needed. However, there are situations where we define a conditional relationship into the question. For example, suppose we want to know what the probability is of the sum of the two dice equaling 5 when we roll the dice sequentially and the first die gives us 3 or less.

Let A be the first event, we roll one die and get 3 or less. Since getting 3 or less must happen half the time, this probability is .5. Let event B be the second event, we roll the second die and get a total of 5.

$P(A \text{ and } B)$ is the probability of getting a total of 5 from rolling two dice while the first die is 3 or less. The possibilities are {1,4}, {2,3}, and {3,2}, giving us a probability of 3/36.

The probability of getting a total of 5 when the first die gave us 3 or less is therefore

$$P(B|A) = \frac{P(A \text{ and } B)}{P(A)} = \frac{\dfrac{3}{36}}{\dfrac{1}{2}} = \frac{1}{6} \qquad (10.12)$$

For simple problems it is often easiest to list the possibilities, show the reduced event space, and get the results.

Suppose we have a bag with 2 red marbles and 2 green marbles. We want to draw them out, one at a time. Since there are 4 choices for the first draw, 3 choices for the second draw, 2 choices for the third draw, and 1 remaining choice for the fourth draw, there are 4! = 24 possible combinations, as shown. Of these 24 choices, 4 of them have the two red marbles being chosen first. The probability of doing this is therefore 4/24 = 1/6.

We'd like to know the probability of drawing a second red marble once we've drawn a red marble. If A is the event of drawing a red marble first and B the event of drawing the second red marble once we've drawn a red marble, then we're looking for $P(B|A)$.

From above, $P(AB) = 1/6$. Since there is an equal number of red and green marbles before we start, the probability of drawing a red marble first is ½. $P(A)$, therefore is ½. Therefore,

$$P(B|A) = \frac{P(AB)}{P(A)} = \frac{\dfrac{1}{6}}{\dfrac{1}{2}} = \frac{1}{3} \qquad (10.13)$$

Doing this problem "manually," the probability of drawing a red marble first is ½. Once we've drawn a red marble, there is 1 red marble and 2 green marbles left (the reduced event space), so the probability of drawing the second red marble is simply 1/3.

Medical Test Results

A somewhat complicated but very important problem[2] is the medical diagnosis problem:

There is a deadly disease that afflicts .1% of the population (one person in a thousand). It is very important to determine if you have this disease before symptoms present themselves. Fortunately, there is a diagnostic test available. However, this test is not perfect. It has two failings. The test will only correctly identify someone with the disease (give a positive result) 98% of the time. Also, the test will give a false positive result (identify the disease in someone who doesn't have it) 2% of the time. You take the test and the results come back positive. What's the probability that you actually have the disease?

The answer to this question is important in cases where there is a vaccine available, but it has side effects. Let's say the vaccine causes a deadly reaction in 15% of the people who take it. If you really have a deadly disease, then a vaccine that could cure you is worth the 15% risk. On the other hand, if you really don't have the disease, then why take any risk with a vaccine? You would like to know what the probability of you having this disease is.

Let's start by defining the two events at hand:

1) D is the event that you have the disease;
2) T is the event that you test positive for the disease.

In terms of the information given above, we know that

1) $P(D)$ = the probability, before you take the test, that you have the disease = .001;
2) $P(T|D)$ = the probability of a positive test given that you have the disease = .98;
3) $P(T|\text{not } D)$ = the probability of a positive test given if do not have the disease = .02.

We want to calculate two numbers:

1) $P(D|T)$ = the probability that you have the disease given that the test result was positive;
2) $P(\text{not } D|\text{not } T)$ = the probability that you don't have the disease given that the test result was negative.

---

2 In the next chapter the complexity will be mostly taken away.

We know that

$$P(D|T) = \frac{P(D \text{ and } T)}{P(T)} \tag{10.14}$$

We can calculate the numerator of Eq. (10.14) directly from the known information,

$$P(D \text{ and } T) = P(D|T)P(T) = (.01)(.98) = .0098 \tag{10.15}$$

Finding $P(T)$ requires a little manipulation. Calculate the intermediate number

$$P(\text{not } D \text{ and } T) = P(T|\text{not } D)P(\text{not } D) = P(T|\text{no } D)(1 - P(D))$$
$$= (.02)(1 - .001) = .01998 \tag{10.16}$$

and then

$$P(T) = P(D \text{ and } T) + P(\text{not } D \text{ and } T) = .0098 + .01998 = .02096 \tag{10.17}$$

And finally,

$$P(D|T) = \frac{P(D \text{ and } T)}{P(T)} = \frac{.00098}{.02096} = .0468 \tag{10.18}$$

You have less than a 5% probability of actually having the disease. Intuitively, this doesn't seem reasonable – how can you test positive on a test that "catches" the disease 98% of the time and still have less than a 5% probability of actually having this disease?

The calculation is correct. The intuition issue comes from how we stated the information above. We left out a critical factor – the background occurrence of the disease. Let's state things a little differently, and also show a very quick and dirty way to estimate the answer.

On the average, ten out of ten thousand people will have the disease (.1% of $10\,000$ = 10) while out of the same $10\,000$ people about 200 people (2% of $10\,000$) will test positive for the disease. The probability of someone who tests positive actually having the disease is therefore about $10/200$ = 5%. This is an approximation because this simple estimate does not correctly take into account the overlap of people who test positive and those who actually have the disease. Correctly handling the overlap lowers the probability a bit from the estimate, as Eq. (10.18) shows.

Armed with this probability, you are now in a much better position to decide whether or not to take a vaccine that has a 15% probability of killing you.

Figure 10.1 shows the probability that you have disease, given a positive result on this same test, versus the background incidence of the disease, $P(D)$. Note that the horizontal axis in this figure is different from the horizontal axis

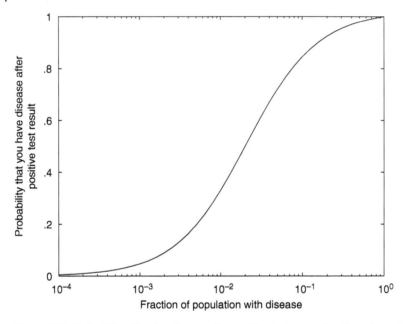

Figure 10.1 Probability of having disease after positive test versus fraction of population with disease.

in previous figures. It is not laid out like a ruler. That is, the distance from .1 to .2 is not the same as the distance from .3 to .4, etc. Instead the spacing is *logarithmic*; there is equal spacing for multiples of ten: the distance from .01 to .10 is the same as the distance from .10 to 1.0, etc. This is done here to make the graph easier to study. If this axis had been laid out with equal distance for equal numbers (call *linear* spacing), then the information for the range below $x = .001$ would have been squeezed up against the left vertical axis and would have been very hard to examine. Since the vertical axis spacings are linear and the horizontal axis spacings are logarithmic, this type of graph is called a *semi-logarithmic* graph.

From Figure 10.1, you can see that at the extremes, it almost doesn't matter whether or not you took the test. If almost nobody has the disease (the left hand side of the graph), then it's very unlikely that you have the disease. An essentially flawless test is required to give you useful information. At the other extreme, if almost everybody has the disease then you probably also have the disease. The test here has a high probability of being correct, but the information you gain doesn't add much to what you already know. For the ranges in between the two extremes, the test gets more accurate as the background (general public) incidence of the disease increases. If 1% of the public has the disease, the test is useful. If 10% of the public has the disease, the test is very accurate.

Returning to the example, we still need to calculate $P(not\ D|Not\ T)$, the probability that you don't have the disease given that the test result was negative. This is simpler than the previous calculation because many of the intermediate numbers are already available. First,

$$P(not\ T) = 1 - P(T) = .97903 \tag{10.19}$$

and

$$P(not\ D) = 1 - P(D) = .99900 \tag{10.20}$$

Then

$$P(not\ D\ and\ not\ T) = P(not\ D) - P(not\ D\ and\ T)$$
$$= .99900 - .01998 = .97902 \tag{10.21}$$

and finally

$$P(not\ D|Not\ T) = \frac{P(not\ D\ and\ Not\ T)}{P(not\ T)} = \frac{.97902}{.97903} = .99999 \tag{10.22}$$

This is as good as it gets. If the test came back negative, you don't have the disease.

The Shared Birthday Problem

The last example on the topic of conditional probabilities is one of the most popular probability book examples: Given a group of n people, what is the probability that at least two of these people share a birthday? Consider only 365-day years and a group with randomly distributed birthdays (no identical quadruplets and the like). We could do this by formal conditional probability calculations, but this problem – like so many popular probability brain-teaser problems – is best solved by a trick. In this case it is much easier to solve the problem of the probability that in a group of n people no two people share a birthday. Then all we need is to subtract this result from 1. This type of calculation was used earlier to calculate the probability of repeated digits in a group of 10 randomly chosen digits.

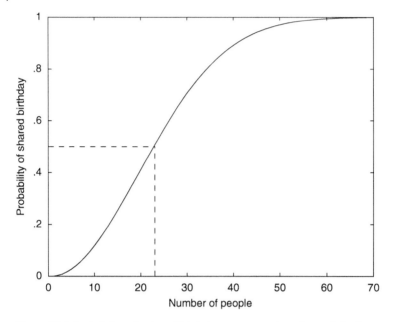

Figure 10.2 Probability of a shared birthday versus number of people in the group.

Pick the first person in the group, an arbitrary choice, and note his/her birthday. Then pick the second person, again an arbitrary choice. The probability that this latter person's birthday is not on the same date as the first person's birthday is $\frac{364}{365} = .99726$. The probability that both of these people share a birthday is therefore $1 - .99726 = .00274$.

Moving on to a third person, if the first two people don't share a birthday, there are two days on the calendar marked off. The probability that this third person doesn't share a birthday with either of the first two is therefore $\frac{363}{365}$ and the probability that there is no match between any of the three is $\frac{364}{365}\frac{363}{365} = .99180$. The probability that at least two of these three people share a birthday is $1 - .99180 = .0082$.

Continuing on (this is easy to do on a spreadsheet or with a hand calculator) is straightforward. Keep repeating the multiplication of decreasing fractions until you've included as many people as you want, then subtract the result from 1.

Figure 10.2 shows the result of these calculations for up to 70 people. The probability of at least two people sharing a birthday crosses 50% at 23 people and exceeds 99.9% at 70 people.

Problems

10.1 This problem and similar variations comes up often on the internet and in texts. It facilitates a good discussion of independent and conditional events.

We have a weighted coin with $P(head) = \frac{3}{5}$. It is flipped three times.

A What is the probability that we get heads exactly two of the three flips?

B What is the probability that we get heads on exactly two of the three flips given that the first flip comes up tails?

C Are the events of getting heads on exactly two of the three flips and getting tails on the first flip independent?

10.2 Figure P10.2.1 shows the standard electrical circuit diagram for a simple switch. This switch may be either in the upper Open or Off position or in the lower Closed or On position. In the Closed position, current may flow from point a to point b, in the Open position the circuit is "broken" and current may not flow from point a to point b.

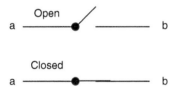

Figure P10.2.1

Figure P10.2.2 shows a circuit made up of 3 switches. The switch positions are random, with a 50% probability that each switch will be Open or Closed. What is the probability that current can flow from point a to point b?

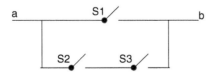

Figure P10.2.2

Notes: In the figure, the switches are labeled S1, S2, and S3 as identification. They are all shown in the Open position; this does not imply anything.

10.3 Repeat Problem 10.2; this time use the circuit of Figure P10.3.1, which shows a circuit made up of 4 switches.

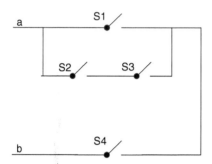

Figure P10.3.1

10.4 A company produces the same product at two different manufacturing plants. Plant 1 produces 25% of the total quantity with a 10% rate of defective parts. Plant 2 produces the remaining 75% with a 5% rate of defective parts. The outputs are combined and then a part is chosen at random. It is defective. What is the probability that it came from each of the plants?

10.5 This is a problem that shows up, in one form or another, in many probability books. The reason it's so popular is that, like the Monty Hall problem, the answers are counterintuitive to many people.

A Jane has three children. At least two of them are boys. What is the probability that the third child is a girl?

B Jane has two children. Both of them are boys. She is now pregnant with a third child. What is the probability that this (third) child will be a girl?

10.6 The Smith family has two children.

A You choose one of the children randomly (without looking). You bet that this child is a boy. What is the probability that you are right?

B Still without looking at the actual children, you are told that the child you did not choose is a boy. Do you want to switch your bet to saying that the child you chose is a girl?

11

Bayesian Statistics

Bayes Theorem

In Chapter 10, we derived and used the following formula for conditional probability:

$$P(D|T) = \frac{P(D \text{ and } T)}{P(T)} \tag{11.1}$$

Rewriting Eq. (11.1) as

$$P(D \text{ and } T) = P(D|T)P(T) \tag{11.2}$$

Since $P(D \text{ and } T)$ must be equal to $P(T \text{ and } D)$,

$$P(D|T)P(T) = P(T|D)P(D) \tag{11.3}$$

from which

$$P(D|T) = \frac{P(T|D)P(D)}{P(T)} \tag{11.4}$$

Equation (11.4) is known as Bayes theorem, after the mathematician who first described it, more than 300 years ago. As we shall soon see, it leads to an entirely different definition of probability and approach to many problems than we have seen thus far.

There is an interesting history of debate between followers of the two approaches. In this book, we shall present both approaches and give examples of their utility and not weigh in on either side of the debate.

Probably Not: Future Prediction Using Probability and Statistical Inference, Second Edition. Lawrence N. Dworsky.
© 2019 John Wiley & Sons, Inc. Published 2019 by John Wiley & Sons, Inc.
Companion website: www.wiley.com/go/probablynot2e

However, rather than beginning with formal definitions, we'll start by using Bayes formula in a few example problems. Along the way, we'll rename the terms in Bayes formula to align with this new way of thinking.

Repeating the medical test problem of Chapter 10: there is a deadly disease that affects .1% of the population. A test for the disease yields 2% false-positive and 2% false-negative results. If you take the test and get a positive result, what's the probability that you have the disease? If you take the test and get a negative result, what's the probability that you don't have the disease?

Use the same nomenclature as in Chapter 10:

D is the event that you have the disease
T is the event that you take the test and get back a positive result
T' is the event that you take the test and get back a negative result

Our first goal is to find $P(D|T)$, the probability that you have the disease given that you took the test and got back a positive result.

Create a table (Table 11.1) in which the columns represent all possible situations. In this example, these are D (you have the disease) and D' (you don't have the disease). The first row of the table will be $P(D)$ and $P(D')$, which we will now call our Prior Information. This is the best information we have about whether or not you have the disease before you take the test. $P(D)$ is given to us as .001; therefore, $P(D') = 1 - P(D) = .999$.

In the second row we enter $P(T|D')$, the probability that the rest returns a positive result given that you don't have a disease, which is given as .02 and $P(T|D)$, the probability that the test returns a positive result given that you do have the disease, which is $1 - .02 = .98$. We'll call this row the Likelihoods. Next, we calculate the numerator of Bayes formula by multiplying the Prior Information and Likelihood numbers.

Table 11.1 Bayes table for medical test, positive test result.

| | Medical test | | |
| | Positive test result | | |
	D	D'	den
Prior information	$P(D) = .001$	$P(D') = .999$	
Likelihood	$P(T|D) = .98$	$P(T|D') = .02$	
Numerator	.00098	.02	.02098
Posterior	.0468	.953	

To finish the calculation we need the denominator of Eq. (11.4), $P(T)$. In Chapter 10, this was an involved calculation.

However, now we are in a better position to get what we need quickly. The probability that you have the disease given a positive test plus the probability that you don't have the disease given a positive test must = 1. This is why we created columns to cover all the possibilities that a positive test result could imply. Therefore, what we need to do is to calculate a normalizing constant, den, which must be $P(T)$: den = .00098 + .02 = .02098. Then we divide both numerator terms by den.

This last row of Table 11.1 shows the results. We'll call this last row posterior probabilities, or more simply, our result. Note that the numbers in the Posterior (bottom) row have been rounded to three significant figures.

$P(D|T)$, our desired result, is .0468. This is of course identical to the result calculated in Chapter 10, but this time there was much less work involved getting it.

The probability that you don't have the disease if you get a negative result, $P(D'|T')$, is easily calculated in a similar manner. Table 11.2 looks the same as Table 11.1, except that the likelihoods now reflect the negative result statistics.

Again, the results repeat those of the earlier calculation. If you take the test and get a negative result, it's almost certain that you don't have the disease.

As an aside, medical test statistics are often stated using a different terminology than was used here. The term *sensitivity* is a measure of false positives: A test with X sensitivity returns $1 - X$ false positives. The term specificity is a measure of false negatives: A test with Y specificity returns $1 - X$ false negatives. In the example above, both the sensitivity and specificity are 98%.

Table 11.2 Bayes table for medical test, negative test result.

	Medical test		
	Negative test result		
	D	*D'*	den
Prior information	.001	.999	
Likelihood	.02	.98	
Numerator	.00002	.97902	.97904
Posterior	.00002	.99998	

Multiple Possibilities

In this section, we'll consider some situations where there are more than two possible outcomes and look at some of the consequences of different choices of priors.

Suppose your friend has a coin that she won't let you examine. You are told that it's either a standard ($HT = 1$ head, 1 tail, evenly balanced) coin or it's a $HH = 2$-headed coin or it's a $TT = 2$-tailed coin. She will flip the coin as many times as you wish and share the results of the flip with you; this will be your only source of information about the true nature of this coin.

You are asked to predict, in terms of probabilities, which of the above three coin types this coin is.

What do you know before the coin is flipped? Very little. Useful information would be the probability of each of the three possible coin types. But you don't know that. Since you have to start somewhere, it seems reasonable to say that each of these three coin types is equally probable, i.e. there's a 1/3 probability that it's an *HH*, a 1/3 probability that it's a *TT*, and a 1/3 probability that it's an *HT* coin. Remember that these are just guesses. We hope they're reasonable guesses, but who knows?

The above probabilities are our prior assumptions for this problem. Priors can be based on information or, as in this case, a subjective guess based on a lack of better information. The choice of priors will prove to be a philosophical sticking point.

We will calculate the probability that we have an *HH* coin if we flip the coin and get a head. Our prior assumptions are that $P(HH) = P(HT) = P(TH) = \frac{1}{3}$.

Let *H* be the event that we flip the coin and get a head. The three likelihoods of getting a head given each of the three coin types are

1) The likelihood of getting a head when flipping an *HH* coin, $P(H|HH) = 1$
2) The likelihood of getting a head when flipping an *HT* coin, $P(H|HT) = \frac{1}{2}$
3) The likelihood of getting a head when flipping a *TT* coin, $P(H|TT) = 0$

We build Table 11.3, as in the previous example. In this case, we have columns for all (three) coin types. Remember than we have to enumerate every possible result I order that our simple calculation of the denominator will be correct.

The posterior case for a *TT* coin when the first flip yielded a head must be 0. This is common sense; there is no way to flip a *TT* coin that yields anything other than a tail. This column could be eliminated almost immediately from the table, but is shown here to keep the example complete.

The posterior row above tells us that we believe the probability of the coin being *HH* is 2/3 and of it being *HT* is 1/3. In other words, we're more convinced that it's an *HH* coin than that it is an *HT* coin, but not overwhelmingly so. This makes sense – we don't have very much data from which to draw a strong conclusion.

Table 11.3 Bayes table for flipping coin with *HH*, *HT*, or *TT* possibilities, one flip (*H*).

| | Coin flip | | | |
| | One flip: Head | | | |
	HH	HT	TT	den
Prior information	1/3	1/3	1/3	
Likelihood	1	1/2	0	
Numerator	1/3	1/6	0	1/2
Posterior	2/3	1/3	0	

Table 11.4 Bayes table for flipping coin with *HH*, *HT*, or *TT* possibilities, two flips (*H, H*).

Two coin flips – two heads				
	HH	HT	TT	den
Prior information	1/3	1/3	1/3	
Likelihood	1	1/2	0	
Numerator 1	1/3	1/6	0	1/2
Posterior 1	2/3	1/3	0	
Numerator 2	2/3	1/6	0	5/6
Posterior 2	4/5	1/5	0	

Suppose we flip the coin again and get a second head. We can update our calculation by just doing it again. In terms of Table 11.4, our previous posterior is now our new prior, and the likelihoods have not changed so there is no need to rewrite them.

We now have an 80% probability that we have a two-headed (*HH*) coin.

What if the second flip had returned a tail?

Table 11.5 shows this situation. In this table, both likelihoods have been written out because they are not the same (as each other). If the second flip had yielded a *T*, then the *HH* coin probability becomes zero and the probability that we have an *HT* coin becomes 1, a certainty. Furthermore, once we have seen both a head and a tail result, we can continue flipping forever and the conclusion that we have an *HT* coin would not change.

This latter observation is supported by common sense; once we've seen both a Head and a Tail, we've narrowed our choices down to just the *HT* coin. Also, note that the order of occurrences doesn't matter. Both pieces of new

Table 11.5 Bayes table for flipping coin with *HH*, *HT*, or *TT* possibilities, two flips (*H*, *T*).

Two coin flips – one head and one tail

	HH	HT	TT	den
Prior information	1/3	1/3	1/3	
Likelihood (*H*)	1	1/2	0	
Numerator 1	1/3	1/6	0	1/2
Posterior 1	2/3	1/3	0	
Likeliood (*T*)	0	1/2	1	
Numerator 2	0	1/6	0	1/6
Posterior 2	0	1	0	

Table 11.6 Bayes table for flipping coin with *HH*, *HT*, or *TT* possibilities, three flips (*H*, *H*, *H*).

Three coin flips – three heads

	HH	HT	TT	den
Prior information	1/3	1/3	1/3	
Likelihood	1	1/2	0	
Numerator 1	1/3	1/6	0	1/2
Posterior 1	2/3	1/3	0	
Numerator 2	2/3	1/6	0	5/6
Posterior 2	4/5	1/5	0	
Numerator 3	4/5	1/10	0	9/10
Posterior 3	8/9	1/9	0	

information (the results of the flips) add the same amount of information and either can come first without affecting the final result.

Returning to the repeated *H* flips situation, after three flips we have Table 11.6. The more times we flip this coin and get a head, the more convinced we are that this is an *HH* coin and that we will never see a tail.

Suppose that initially your friend had told you "I think it's a normal (HT) coin" which you interpreted to mean that there's a 50% probability that it is indeed an *HT* coin, a 25% probability that it's an *HH* coin, and a 25% probability that it's a *TT* coin. These numbers are now the first-flip priors. Repeating the above table, we get Table 11.7. The original assumptions (priors) have a strong effect early on, but their effect diminishes with each new set of data.

Table 11.7 Bayes table for flipping coin with *HH, HT,* or *TT* possibilities, three flips (*H, H, H*), revised priors.

Three coin flips – three heads, revised priors

	HH	HT	TT	den
Prior information	.250	.500	.250	
Likelihood	1	1/2	0	
Numerator 1	.250	.250	0	.500
Posterior 1	.500	.500	0	
Numerator 2	.500	.250	0	.750
Posterior 2	.667	.333	0	
Numerator 3	.667	.167	0	.833
Posterior 3	.800	.200	0	

The next example is similar to the previous example. The differences are that the number of possible results has been increased and that these results are described by a simple function rather than a description.

We have a coin which has a Head and a Tail, but it's a weighted coin. We'll describe its probability of yielding a Head when flipped by the number r. For this example, we will allow five values of r:

- $r = 1.0 \rightarrow 100\%$ probability of yielding a head
- $r = .75 \rightarrow 75\%$ probability of yielding a head
- $r = .50 \rightarrow 50\%$ probability of yielding a head
- $r = .25 \rightarrow 25\%$ probability of yielding a head
- $r = .00 \rightarrow 0\%$ probability of yielding a head

Start out by assuming (our prior information) that all five values of r are equally likely.

The likelihoods of a flip yielding a head from each of the coins are each identically r. This is not a coincidence, r was defined so that this would be the case. This is really just an extension of the previous example: $r = 1.0$ was identically the HH case in the previous example, $r = .5$ was the *HT* case, and $r = .0$ was the *TT* case. As a direct extension, the likelihoods of a flip yielding a tail are $1 - r$.

Table 11.8 shows the results of flipping three heads in a row. Both the *H* and *T* likelihoods are shown so that they may be applied conveniently when/ if needed.

With each succeeding flip that yields a head, we are more convinced that r is 1. We know for certain that r is not 0 as soon as we have seen a head. However, we are never fully certain that r cannot be any of the intermediate values.

Table 11.8 Bayes table for flipping coin with multiple possible weightings (H, H, H).

	$r = 1.0$	.75	.50	.25	.00	den
Prior information	.20	.20	.20	.20	.20	
H Likelihoods	1.0	.75	.50	.25	.00	
T Likelihoods	.00	.25	.50	.75	1.00	
Flip → H, Nums	.200	.150	.100	.050	.000	.500
Posterior 1	.40	.30	.20	.10	.00	
Flip → H, Nums	.400	.225	.100	.025	.00	.750
Posterior 2	.533	.300	.133	.033		
Flip → H, Nums	.533	.225	.0667	.0083	.00	.833
Posterior 3	.64	.27	.08	.01	.00	

What if we asked somewhat different questions about our results. Our available values of r allow for a coin that either favors heads, is neutral, or favors tails. What are these probabilities? After seeing three heads in a row and no tails (Table 11.8), our predictions are

1) .01 probability that the coin favors tails.
2) .08 probability that the coin is neutral (what we've been calling an HT coin).
3) $.64 + .27 = .91$ probability that the coin favors heads.

The wording of the question must be read very carefully. For example, if the question was "What is the probability that the coin does not favor tails?," our answer based on these same 3 flips would be $.64 + .27 + .08 = .99$.

Returning to the coin flips, what if one of the flips yields a tail? We can change any of the flips in the table to reflect this, it doesn't matter which flip we change. In this example we'll change the last (third) flip (Table 11.9).

Having seen both a Head and a Tail, we're sure that r can be neither 0 or 1. Based on having seen two Heads and only one Tail, we think it's most probable that $r = .75$. However, because of our limited information, we don't see $r = .75$ as much more probable than $r = .50$. This process can be continued indefinitely, with the results reflecting the results of the flips we get.

Finally, let's return to the case of 3 flips yielding 3 heads, but with an added twist: The light in the room is poor and we can't see the results of the flips too clearly; we only correctly read the result (of a flip) 80% of the time.

We are still interested in the probabilities of the values of r, but our estimate can only be based on what we see, not on the actual result of the flip. The likelihood calculations of the previous examples must be revised.

Table 11.9 Bayes table for flipping coin with multiple possible weightings (H, H, T).

	r = 1.0	.75	.50	.25	.00	den
Prior information	.20	.20	.20	.20	.20	
H Likelihoods	1.0	.75	.50	.25	.00	
T Likelihoods	.00	.25	.50	.75	1.00	
Flip → H, Nums	.200	.150	.100	.050	.000	.500
Posterior 1	.40	.30	.20	.10	.00	
Flip → H, Nums	.400	.225	.100	.025	.00	.750
Posterior 2	.533	.300	.133	.033		
Flip → T, Nums	.000	.075	.0667	.025	.00	.167
Posterior 3	.00	.45	.40	.15	.00	

If we have a coin with $r = 1$, it is certain that a flip will yield a head. But, we'll only see a head 80% of the time, so $P(H|r = = 1) = .8$. If we have a coin with $r = 0$, it is certain that a flip will yield a tail. But, we'll only see a tail 80% of the time, so $P(H|r = = 1) = .2$. Generalizing, our likelihoods may all be calculated from the formula

$$P(H|r) = .8P(H) + .2P(T) = .8r + .2(1-r) = .2 + .6r \qquad (11.5)$$

Revising the 3 flips yields 2 heads and 1 tail in Table 11.9, we get Table 11.10. After flipping two heads and one tail, our best guess is that $r = .75$. This is not a strong guess, it's hardly a higher probability than $r = 1.0$ or $r = .5$. However, neither $r = 1.0$ nor $r = .0$ can be ruled out, because of the uncertainty in seeing the results of the flips.

Will Monty Hall Ever Go Away?

Let's reexamine the Monty Hall problem using the Bayes Theorem formalities of this chapter. Refer back to Chapter 1 for the description of the problem. We'll call the three doors A, B, and C, with the prize behind one, and only one, of these doors. Events A, B, and C are the prize being behind door A, B, or C, respectively.

The prior probabilities are known in this case. There's an equal chance of the prize being behind any of the doors. In our notation, $P(A) = P(B) = P(C) = \frac{1}{3}$.

Let's choose door A (an arbitrary choice; remember, none of the doors are open).

Table 11.10 Bayes table for flipping coin with multiple possible weightings (*H, H, T*), poor visibility.

	r = 1.0	.75	.50	.25	.00	den
Prior information	.20	.20	.20	.20	.20	
H Likelihoods	.80	.65	.50	.35	.20	
T Likelihoods	.20	.35	.50	.65	.80	
Flip → *H*, Nums	.160	.013	.100	.070	.040	.500
Posterior 1	.32	.26	.20	.14	.08	
Flip → *H*, Nums	.26	.13	.10	.07	.04	.50
Posterior 2	.38	.29	.17	.08	.03	
Flip → *T*, Nums	.09	.10	.09	.05	.02	.35
Posterior 3	.25	.29	.24	.1	.06	

The host must now open either door *B* or door *C*, but the prize cannot be behind the door he opens (he can do this because he knows where the prize is). Let's say he opens door *B*. Of course, the car is not there. We now know that the car is either behind door *A* or door *C*. We must decide to stay with our choice of door *A* or switch to door *C*.

For our table, we need the likelihoods of the events *A*, *B*, and *C*:

$P(pickB|A)$= probability that the host would open door *B* if the prize is behind door *A*. From the host's point of view, knowing that the prize is behind door *A* and that he is allowed to choose either doors *B* or *C*, he will randomly pick. Therefore, the probability that he will choose door *A* = .5.

$P(pickB|B)$= probability that the host would open door *B* if the prize is behind door *B*. The host knows he is not allowed to show the prize. This is not an allowed choice, so the probability = 0.

$P(pickB|C)$= probability that the host would open door *B* if the prize is behind door *C*. If the prize is behind door *C* (again, the host knows this), his only choice is door *C*, so this probability = 1.0.

Table 11.11 shows all the information.

Reviewing, our original guess for the prize had been door *A*. The host's available constraints and choices are summarized in the Likelihoods. The Posterior row shows that the probability that the prize is behind door *A* is 1/3, that it is behind door *B* is 0, and that it is behind door *C* is 2/3. Clearly, we want to use our option to switch to the unopened door, door *C*.

The simplicity of Table 11.11, especially as contrasted to the myriad of discussions about the "right answer" to the Monty Hall problem, is striking.

Table 11.11 Bayes table for the monty hall problem.

	A	*B*	*C*	den
Priors	1/3	1/3	1/3	
Likelihoods	1/2	.0	1	
Numerators	1/6	.0	1/3	1/2
Posteriors	1/3	.0	2/3	

Philosophy

Bayes formula was introduced in Chapter 10. We might therefore conclude that "there is really nothing new in this chapter, it's just a few interesting examples." In fact, the way the examples in this chapter are dealt with implies a very different meaning of *probability*.

If we believe, for example, that the probability of a commercial airliner crashing is a fixed number, we are saying that there is a probability to the long-run frequency of events. This approach is called the Frequentist description of probability. We often broaden this definition to include "parallel universes." That is, we note that flipping 1000 identical coins once each is the same as flipping 1 coin 1000 times – insofar as the probability of getting a certain number of heads is concerned.

The Frequentist approach is self-consistent, but often leads to conclusions which don't look as good "on the street" as they do on paper.

For example, if I have a coin that is exactly, or even approximately, our ideal HT coin, then it doesn't matter how many times I flip it or what results I get, the probability of a head on the next flip is always ½.

But what if we revise our meaning of probability to be a measure of what we believe the result of (in this example) the next coin flip will be? This is called the *Bayesian*, or diachronic interpretation. Diachronic in this usage means "evolving over time," or "evolving as a result of new information."

The coin flip examples in this chapter illustrate this way of thinking. We start out with a coin that we know nothing about. We can formalize the fact that we know nothing by assigning a uniform probability density to the coin's propensity to flip to a head (we'll do this formally in the next section). Or, if we have some reason to believe that this is a true *HT* coin, we can assign a high probability to what was called $r = .5$ in previous sections and a lower probability to other values of r. We are making a prediction based upon our beginning knowledge, or lack of knowledge, as the case may be.

Then we flip the coin; either just one time or many times. With each succeeding flip we incorporate the result into our prediction and update the prediction. If you flip a coin ten times and get ten heads, just how willing are you to believe (without additional information) that this is an *HT* coin? Would you take the very naïve approach that since ten heads and one tail are more probable for eleven flips of an *HT* coin than are eleven heads, a tail "is due" and bet on it? Or would you conclude that a coin that just gave you ten heads is weighted to give heads, and the eleventh flip will most likely be a head? This latter conclusion, which seems to satisfy common sense and experience best, is exactly what the Bayesian calculation predicts.

The Prosecutor's Fallacy

Bayesian calculations can lead to a very clear description of "the Prosecutor's fallacy." This is best introduced by an example:

Suppose you live in a city of 1 million people. You are over 6'1" tall (as is about 1% if the population), you are male (as is about 50% of the population), you are left-handed (as is about 10% of the population), and you have unusual hazel-colored eyes (as is about 5% of the population).

If we assume that these traits are independent, then about $(.01)(.5)(.1)$ $(.05) = .000025 = 25$ out of a million people fit your description.

A robbery was committed on a busy city street, and numerous witnesses identified a very tall man with hazel-colored eyes, holding a gun in his left hand. The next day a policeman sees you on the street and arrests you for this robbery.

In court, the prosecutor makes his case as "Only 25 out of a million people match this description. This means that the probability of this defendant's guilt is $1 - \frac{25}{1000000} = .999975$. This meets the definition of "beyond a reasonable doubt."

Really?

The fallacy can be seen informally if you use all the information given, rather than just some of it: In a city of one million people, there are about 25 people meeting this description. Without any more evidence pointing to you, the approximate probability of your guilt is just $\frac{1}{25} = .04$, hardly a probability of guilt "beyond a reasonable doubt."

What the prosecutor has calculated is P(Fits the Description | Guilty) when he needs to calculate P(Guilty | Fits the Description). In other words, he needs Bayes Theorem.

Let's begin with a simple prior. In a city of a million people, someone chosen at random has a probability of 1 in a million of being guilty.[1] The probability of this person being not guilty is .999999.

---

1 Since we're not including the victim in the list of possible robbers, this should actually be 1 in 999 999. But, we'll allow a little sloppiness here.

Table 11.12 Bayes table for the prosecutor's fallacy.

	Guilty	Not guilty	den
Priors	.000001	.999999	
Likelihoods	.999975	.000025	
Numerators	.000001	.000025	.000026
Posteriors	.038	.961	

The likelihoods are what the prosecutor has been throwing around. The likelihood of guilt is $1 - \frac{25}{1000000} = .999975$. The likelihood of innocence, or "not guilt" is $\frac{25}{1000000} = .000025$.

Now we build our usual table (Table 11.12). This table shows a probability of guilt of just over 3.8%, very close to the 4% of the "quick and dirty" calculation shown above. The difference is again due to double counting.

Analogous to the Prosecutor's fallacy there is of course a Defendant's fallacy, and an untold number of variations on this theme.

Continuous Functions

The examples in this section require the use of integrals to normalize the Bayes Theorem Calculation. For readers unfamiliar with basic integral calculus, just accept that we are normalizing the numerator to get the posterior, just as in the previous examples. Reading the Appendix will help.

In the coin flip example above, let us expand the number of allowed values of r from 5, to 25, to 250, to ..., all the way up to a continuous function. We'll stay with the same coin flip example and choose the function for the probability of flipping a head to be

$$f(r) = r, \quad 0 \le r \le 1 \tag{11.6}$$

Because of our choice of function, we have a simple correspondence to the previous examples; $r = 1$ is an *HH* coin, etc. Note that, as in previous comparisons of discrete and continuous probability distribution functions (PDFs), we have to pay attention to what it means to pick a value (discrete distribution) or pick a range (continuous function).

Our likelihood of flipping a head is simply r. The likelihood of flipping a tail is $1 - r$.

We no longer need to write out a table; the table would have only one column.

A given coin must have a certain value of r, depending on how it's weighted; it does not have a distribution function. However, just as we calculated posterior probabilities for the possible values of r, now we calculate a function, $p(r)$, for the posterior probabilities.

Assuming a uniform probability density as our prior distribution, after one flip that yields a head,

$$p(r) = \frac{(1)(r)}{den} \tag{11.7}$$

The denominator in the above expression, den, is the normalizing constant that makes $p(r)$ a PDF. For continuous functions, this is the integral over all r. Since the numerator is only nonzero between 0 and 1, this is all we need to consider:

$$den = \int_0^1 r\,dr = \frac{1}{2} \tag{11.8}$$

And therefore

$$p(r) = 2r \tag{11.9}$$

Based on one flip of a head, we predict that the probability that the coin is weighted toward heads is

$$p(headweighting) = \int_{\frac{1}{2}}^1 2r\,dr = .75$$

We can of course calculate the probability for any region of r.

In this example, calculating den was a simple exercise. Not only will this not always be true, but there will be cases where the required integral cannot be formally evaluated. However, den is a number, not a function; a numerical evaluation is always satisfactory.

As in the discrete case, we can continue flipping the coin and updating our calculation; each new prior is the previous posterior.

Figure 11.1 shows $p(r)$ for one, two, and three flips that yielded heads. As would be expected, we are getting more confident in higher values of r and less confident in lower values of r. After 3 successive heads, we are approximately 94% confident that the coin is biased (weighted) toward heads.

In Figure 11.2, we (re)show the result for 3 Heads and add the result for 3 Heads + 1 Tail and then 3 Heads + 2 Tails. The latter curve is beginning to resemble a Gaussian curve.

The curves in Figures 11.1 and 11.2 were generated by repeatedly invoking Bayes Theorem, starting with a uniform prior, using r as the heads likelihood or $(1 - r)$ as the tails likelihood, and building up the desired posteriors. We can

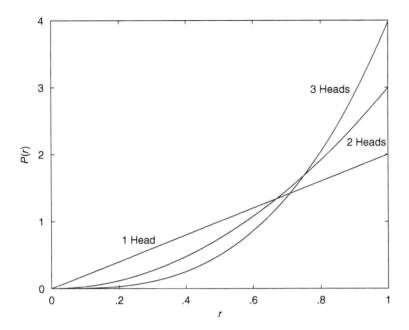

Figure 11.1 Probability of weighting (*r*) versus *r* for *H, HH*, and *HHH* flips.

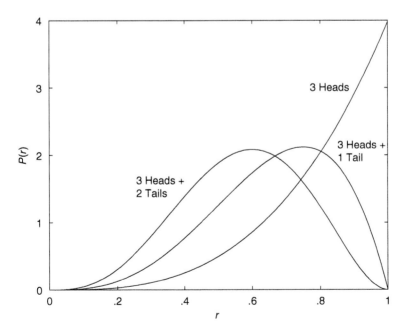

Figure 11.2 Probability of weighting (*r*) versus *r* for *HHHH, HHHT*, and *HHHTT* flips.

get the same results without the iterative process by invoking the binomial probability formula that has been described in Chapter 6. Since we will be normalizing to get the posterior function, we can use the binomial formula without its normalizing component as our likelihood function

$$Likelihood = r^k (1-r)^{n-k} \tag{11.10}$$

where k is the number of head and n is the total number of flips $(n + k)$ and we use a uniform priority function.

There is no need to show graphs of the results of the above procedure; they are identical to Figures 11.1 and 11.2.

For a more realistic situation, assume that we are in the coin minting business. Our ideal minting machine would produce coins with exactly $r = .5$. However, due to machine tolerances and material nonuniformities, the best we can get is a machine which produces a narrow distribution of r, centered at $r = .5$.

An example of such a distribution is a normal distribution with $\mu = .5$ and $\sigma = .1$, as shown in Figure 11.3. In a real factory we'd probably be unhappy with this large a standard deviation, but we'll use this number for the sake of the example.[2] We will redo the above example using this distribution as our prior.

In the same figure, we show the posterior distributions after flipping one, two, and then three heads. As the figure shows, with each succeeding head the peak of the curve (most likely value) moves in the $+r$ (to the right) direction.

If we were to add some Tail flips, the peak would move to the left, as expected.

Credible Intervals

Figure 11.3 contains a lot of information. We need to summarize this information somehow to make it useful. This was done informally by noting the direction of the movement of the peaks of the curves. Tabulating these results would be our first summary information from the analysis. The fraction of the curve to the right of $r = .5$, indicating the probability that the coin is weighted toward heads, would be our second summary information.

Another summary we can calculate is analogous to the confidence interval used for Gaussian curves. Since we are looking at curves that resemble Gaussian distribution curves, it is easy to intuitively picture what we want. However,

---

2 Since a Normal distribution extends to infinity in both directions on the horizontal axis, we are opening the door to negative values of r which makes no physical sense. However, since in this example $r = 0$ is five standard deviations below the mean, this probability is so low that we will not worry about it. There are of course many functions which would not present this problem but the Gaussian function has been examined in earlier chapters and introducing new functions here is an unnecessary complication.

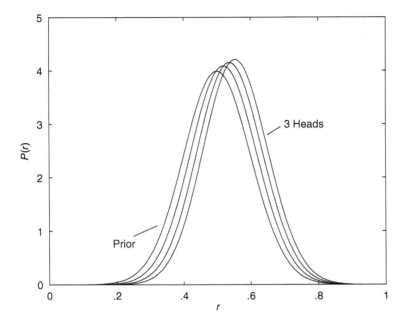

Figure 11.3 PDF of coin minting machine output versus number of *H* flips.

since these curves are not really Gaussian curves (except for the original prior curve), we cannot derive a general formula; we will need to perform the calculation numerically.

We will call these new number Credible Intervals rather than Confidence Intervals. They can be calculated for any curve width we want: 2.5–97.5% is typical, reminding us of the two-sigma confidence intervals for Gaussian curves. If our curves happen to be Gaussian, the Credible and the Confidence Intervals are identical.

Referring again for Figure 11.3, the 2.5–97.5% (95% wide) credible interval for our prior function is .400. This is not a surprise, this is 4σ for a normal function. The same credible interval for the three heads curve is .375. In other words, as we get more information, not only do we update our estimate of μ, our credible interval about μ gets narrower.

Gantt Charts (Again)

Thinking as a Bayesian rather than as a Frequentist, let's re-examine the discussion of Gantt charts that was presented in Chapter 10.

Probability predictions of the completion date of a project and, by implication, the capabilities of the manager in charge, may be calculated in the same manner as calculating the probability of the results of flipping a weighted coin. Actually, the calculations are identical – we just have to rename the terms.

Consider r to be the probability that a manager will get a project finished on or before schedule. r will take on values from 0 to 1. $r = 1$ means that it's certain that the project will be finished on or ahead of schedule, analogous to $r = 1$ meaning it's certain that the flip of a weighted coin will yield a head. $r = 0$ means that it's certain that the project will finish late, analogous to $r = 0$ meaning that it's certain the coin flip will not yield a head. $r = .5$ means that it's just as likely that the project will finish early as it is that the project will finish late, analogous to $r = .5$ meaning that it's equally probable that the coin flip will yield a head or a tail.

Now we simply return to the continuous function weighted coin example's results. Instead of flipping a coin and getting a Head, we run a project and get an Early, similarly for Tail and Late.

The analysis parallels the explanation of the weighted coin flip analysis. This is not a surprise because it's the same analysis, relabeled. We could (and in the real world should) do a more sophisticated analysis in which we include how early or late the project is finished, some managerial oversight weighting factor for how easy or difficult the project is, etc. Here we will keep it simple and just use the Heads = Early, etc., analogies.

$P(r)$ is the PDF that the project will finish early versus r. Since we are looking at continuous functions, we must consider nonzero interval widths of r to evaluate probabilities.

The information needed for a managerial decision is contained in these curves, but it must be extracted with further calculations. Management wants to know what the probability is of bringing a project in early, based on a project leader's prior record. In order to extract this information from these curves, we need to calculate the expected values of the PDFs,

$$EV = \int_{0}^{1} rp(r)\,dr \qquad (11.11)$$

and the credibility intervals for several cases.

Without showing the details of the calculations (again, they parallel the coin flip calculations), Table 11.13 shows the results of several calculations.

Referring to Table 11.13, the "3 Early" result shows a very high EV of bringing the project in early, as we would expect. A Credibility Interval for the result is not shown because this curve has a different shape than the others (see Figure 11.2) and a direct comparison is not possible. The rest of the results in the table show EVs as would be expected. However, note that the Credibility Interval does not significantly shrink until there is a lot of data.

As suggested above, this example would be more realistic if other factors were brought into the data because all projects are not equally difficult. However, it does show that Bayesian calculations can be useful and not difficult to interpret for Gantt Charts and the like.

Table 11.13 Bayes table Gantt chart interpretation.

History	EV	Credibility interval
None	.50	n.a.
3 Early	.80	See text
3 Early, 1 Late	.67	.67
3 Early, 2 Late	.57	.66
3 Early, 3 Late	.50	.64
15 Early, 10 Late	.59	.37

Problems

11.1 Once the Monty Hall problem is organized in a table as was done in this chapter, it is straightforward to consider some variations. For example, suppose there are four doors. Keep all the rules: (first) choose door A and see the host open door B as in the original problem; what are the probabilities of the available alternatives now?

11.2 Extend the Monty Hall Problem to n doors, $n \geq 3$. You should be able to create simple formulas by extending the tables.

11.3 This problem is a continuation of the coin flip discussion. Again, the probability that the coin will yield a head is r, $0 \leq r \leq 1$. We will redefine our basic experiment to consist of a pair of flips. Evaluate several of the basic results, commenting especially on those results that were impossible to obtain with the single flips discussed in the text.

If you have a computer and the requisite programming skills, do the continuous case. If you are working by hand or with a spreadsheet, do the discrete case with $r = 0, .1, .2, ..., .9, 1.0$ allowed.

11.4 **A** Person A tells the truth 75% of the time. A die is rolled and A reports seeing a 6. What is the probability that the result really was a 6?

B Now person A will tell the truth in proportion to the result of the rolling of a die. If the die roll is actually a 6 he will tell the truth. If it is a 5 he will tell the truth 5/6 of the time, etc. The question is the same as above: A reports seeing a 6. What is the probability that it really was a 6?

11.5 The police have arrested a suspect in a robbery and they are 90% certain that they have the right man. However, he has several witnesses who are 75% certain that they saw him someplace far from the scene of the crime at the time of the crime. Assuming that they are all telling the truth and are all credible witnesses, how many of these witnesses do we need before it's more probable that he's innocent than that he's guilty?

11.6 Suppose we have a person accused of a crime and all of the accumulated evidence and witnesses make us 75% confident that she's guilty.

We bring out a polygraph (lie detector) machine that is correct 75% of the time. The polygraph test tells us she's innocent. What is our position now?

11.7 Following the reasoning leading to Eq. (11.5) (flipping coins in dim light), we can also modify the coin weighting likelihood (r) to include a weight W, $0 \le W \le 1$ where $W = 1$ means we are certain that the data is good, and $W = 0$ means we are certain that the data is worthless and shouldn't affect our conclusion(s).

A For the weighted coin flip situation depicted in Table 11.8, construct a Likelihood function that includes the weighting function described above.

B Using the result of part (a), recreate Table 11.7 using $W = .5$.

11.8 This is the first of several "M&M" problems. M&Ms are small coated chocolate candies that are often seen on desktops in jars containing hundreds of them. The coatings are brightly colored in several easily identifiable colors. For the purposes of the problems to follow, these candies could just as easily be colored marbles or any of a number of different items. Over the years, however, colors taken off the market and new colors added to the mix have been popular discussion topics, and the identification of these Bayes statistics problems as the "M&M problems" looks like it will always be with us.

If we're dealing with only two different color M&Ms in the jar, we could talk about chocolate and vanilla cookies in a jar, or men and women in a room, or ... . For whatever reason, the M&M image sticks best in peoples' minds.

In many of these problems, the number of articles being dealt with make a paper and pencil solution awkward. Spreadsheet solutions are very straightforward. Since modern spreadsheets have functions such as the binomial distribution built in, setting up tables using these functions is very easy.

You have a jar with 10 M&Ms, some red and some blue, in it. Your pull out 5, 3 of which are red (and two of which are blue). Estimate the number of each color of M&Ms that were originally in the jar. How good is this estimate?

11.9 This problem is straightforward using a spreadsheet, it's a bit unwieldly to do by hand. Repeat Problem 11.8 with a mix of 20 red and blue M&Ms in a jar. Pick 5, get 3 red (and 2 blue). Estimate the starting mix and a confidence interval for this estimate.

11.10 Repeat Problem 11.8 but this time pull the two blue M&Ms first, then repeat it but pull Red–Blue–Red–Blue–Red. Convince yourself that you'll get the same result regardless of the order of M&M extraction.

11.11 We have two bowls of M&Ms. The first bowl has 200 red and 100 blue M&Ms, the second bowl has 200 red and no blue M&Ms. We pick a bowl at random and extract an M&M. It is red. What is the probability that it came from the first bowl?

11.12 The table below shows the mix of M&Ms as sold before and in 1995 when blue M&Ms were introduced and the mix changed.

You have two jars of M&Ms. One of them is pre-1995, the other is post-1995, but you don't know which is which. Ignoring the obvious observations that the older jar doesn't have any blue M&Ms while the newer jar doesn't have any tan M&Ms (let's say that the jars are opaque), you randomly pick a jar and pull out an M&M. It is brown. What are the probabilities of which jar this brown M&M was pulled from?

Color	Pre-1995	Post-1995
Blue	.00	.24
Brown	.30	.13
Green	.10	.20
Orange	.10	.16
Red	.20	.13
Tan	.10	.00
Yellow	.20	.14

11.13 A polygraph, or lie detector, is a machine that reports whether the person that the machine is connected to is answering questions truthfully or is lying. However, the machine is not perfect. One such machine type reports two types of errors:

1 The subject passes the test even though he's lying 14% of the time.
2 The subject fails the test even though he's telling truth 12% of the time.

In our notation these statements are written as

1 $P(pass \mid lying) = .14$
2 $P(fail \mid truthful) = .12$

If this machine were to be used to test defendants accused of a crime, we would certainly want to know the probability that they're lying if they fail the test (we're assuming that they will answer the question "Did you do it?" with "No").

For priors, assume that 95% of the human race is truthful.

Hint: If you don't know how to begin, go back to the disease test example in this chapter and Chapter 10. This is the same problem with different nomenclature and numbers.

11.14 What fraction of the general population would have to routinely lie for the answer to the previous problem to be 50%?

11.15 Suppose some unfair dice get mixed in with all the fair dice in a gambling establishment. These unfair dice, 5% of the total dice inventory, will roll a certain number 1/3 of the time. We are about to start shooting craps and want to make sure we don't have one of these unfair dice. We pick up a die and roll it, and get a 4. We roll the die again and again get a 4. This is suspicious but not convincing, so we keep rolling this die and keep getting 4. How many 4's in a row do we need to get before we ask to toss out this die?

12

Estimation Problems

The Number of Locomotives Problem

An important use of Bayes Theorem is for estimating parameters in a situation where there is limited information. The locomotive problem is an example of this estimation category of calculations. We are told that a railroad company numbers its locomotives consecutively from 1 to N, where N is the total number of locomotives owned by the company (and presumably they're all out running on tracks somewhere). We see a locomotive with number 60 painted on it. We are asked to estimate how many locomotives this company has.

We have – at this time – very little information to help us create a prior. Therefore, let us start with the simplest prior possible. If N is the total number of locomotives owned by the company, then our prior is a uniform distribution from 1 to N. We'll try to gain some insight into the implications of our choice of N as we go along.

The likelihoods are straightforward. First, if the company has less than 60 locomotives, there is 0 probability that we saw locomotive number 60. If the company has 60 locomotives, then the probability that the one we saw was number 60 is exactly 1/60. If they have 61 locomotives, then the probability that we saw number 60 is 1/61, etc. In general, if the company has N locomotives, then the probability that we saw number 60 (or any particular locomotive) is $1/N$.

Since we are starting with a uniform prior distribution, the posterior curve and the likelihood curve look exactly the same. Figure 12.1 is the posterior curve for $N = 1000$ (an arbitrary choice).

Figure 12.1 shows that the peak of the curve – the most likely case – is at 60 locomotives. However, what is more useful is to calculate the expected value of this posterior distribution, EV = 333.

Three hundred and thirty three passes the common sense test as a reasonable answer, but just how accurate is it? $N = 1000$ was pretty much "pulled out of the air." What if we try some other values of N?

Probably Not: Future Prediction Using Probability and Statistical Inference,
Second Edition. Lawrence N. Dworsky.
© 2019 John Wiley & Sons, Inc. Published 2019 by John Wiley & Sons, Inc.
Companion website: www.wiley.com/go/probablynot2e

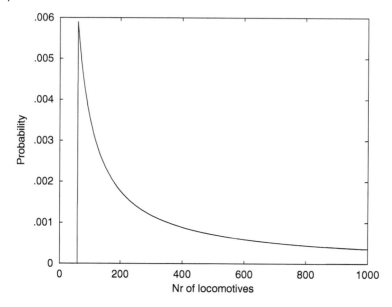

Figure 12.1 PDF of predicted nr of locomotives, $N = 1000$, uniform priors, #60 seen.

Table 12.1 Predicted nr of locomotives versus N, uniform priors, #60 seen.

N	EV	2.5% CDF	97.5% CDF
500	207	62	476
1000	333	63	933
2000	552	64	1833

Table 12.1 shows the expected values of the posterior distribution for several values of N. As can be seen, the expected value is highly dependent on the guess for N. These results are almost useless.

The last two columns in Table 12.1, labeled 2.5% and 97.55% Cumulative Distribution Function (CDF) let us look at credibility intervals. Looking at these entries, we see that these numbers reinforce our conclusion that we do not yet have a good estimate of how many locomotives there are.

Number of Locomotives, Improved Estimate

In order to improve our estimate of the number of locomotives, we need more information. This can be in either or both of two categories: learn more about the statistics of the locomotive business so that we can improve our prior distribution, and spot some more locomotives so that we have more data.

Let us start with the prior distribution. One would think that there cannot be too many railroad lines with thousands of locomotives in service. The probability of a very large number of locomotives must fall with increasing N. On the other hand, the economics of running a railroad line, maintaining a spare parts inventory, etc., say that there will be no – or almost no – railroad lines with very few cars.

We can hypothesize a prior distribution as shown in Figure 12.2, described by the equation

$$prob(N) = Ne^{-\alpha N} \tag{12.1}$$

with $\alpha = .01$.

Table 12.2 is a repeat of Table 12.1, using the prior distribution of Eq. (12.1). Also, the adjustable parameter N has been replaced with the (adjustable) parameter α. Comparing Table 12.2 to Table 12.1, we see that the credible Interval numbers have been greatly improved as compared to the uniform prior results. However, the expected value of the posterior still varies with the choice of an arbitrary parameter. Perhaps if we did a detailed study of the locomotive manufacturers and could come up with a number for α that is not simply a guess, we would be comfortable with the resulting expected value. However, lacking that, we must get more observation data if we want better results.

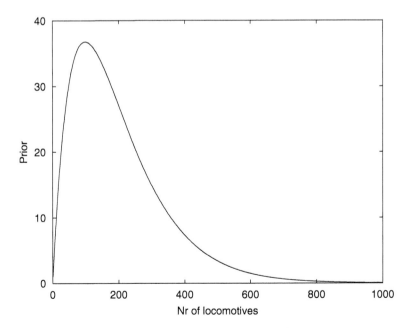

Figure 12.2 PDF of predicted nr of locomotives, improved priors, #60 seen.

Table 12.2 Predicted nr of locomotives, improved priors, #60 seen.

a	EV	2.5% CDF	97.5% CDF
.05	80	59	134
.01	159	61	429
.005	250	64	674

Table 12.3 Predicted nr of locomotives, improved priors, 5 sightings.

a	EV	2.5% CDF	97.5% CDF
.05	112	99	147
.01	127	99	214
.005	133	99	248

Table 12.3 shows what five observations produce. The five observations are 30, 60, 70, 80, and 100. We can now make pretty good guesses at how many locomotives there are and how precise our estimate is.

Before leaving the locomotive problem, let's close the loop on our calculations and gather some statistics on just how good our estimation process really is. Assume a locomotive company with 250 locomotives. We'll generate random numbers in the range 1–250, these will be our sightings. Then we'll run these numbers through the procedure above, fixing a at .01.

Repeat the above procedure 1000 times and generate statistics on how well our process works (i.e. we do a Monte Carlo simulation). We will be generating 1000 expected values and getting their average. This data will look normal, and it makes sense to look at the standard deviation of these (1000) numbers to see the spread of the data (95% credibility, or in this case 95% confidence interval).

Table 12.4 shows the results of this experiment. As the table shows, the number of sightings of locomotive numbers is important to the accuracy of the prediction, but even only a few sightings will give useful information.

Decision Making

In Chapter 11, we looked at weighted coins. We never asked a question such as "If I flip a coin 100 times and get 60 heads and 40 tails, what can I conclude about whether or not this coin is fair?"

Starting with the bottom line, we can never reach a general conclusion with absolute certainty. We deal in probabilities – and as they say – the only

Table 12.4 Monte Carlo simulation of 250 locomotive railroad.

Nr sightings	EV_{avg}	2.5% CDF	97.5% CDF
1	223	83	362
3	243	133	355
5	250	174	326
7	250	191	309

certainties available are death and taxes.[1] We can add to this list ratios – we cannot say what will happen, but we can often compare two or more situations and say which is more likely to happen.

The Bayes calculation of the probability of flipping a coin 100 times and getting 60 heads and 40 tails gives the same result as the binomial distribution of 100 coins yielding 60 heads with the probability p being identically our weighting factor r, assuming a uniform prior for the Bayes calculation. In other words, the binomial probability function – with p as the independent variable – is our likelihood function, as was discussed in Chapter 11.

As an example, suppose we have a coin whose r value is either .5 or .7, but we don't know which. We flip the coin 100 times and get 60 heads. Which coin is it more probable that we have?

Assuming a uniform prior, the probability that we have the $r = .5$ coin is, using the binomial probability function

$$prob(r = .5) = \frac{100!}{(40)!60!}(.5)^{60}(.5)^{40} = .0108 \tag{12.2}$$

And the probability that we have the $r = .7$ coin is

$$prob(r = .7) = \frac{100!}{(40)!60!}(.7)^{60}(.3)^{40} = .0085 \tag{12.3}$$

The ratio of these two probabilities is 1.27:1. Since the ratio is larger than one, it is more probable that we have the $r = .5$ coin.

This ratio is close enough to 1 that while the $r = .5$ coin is most probable, the difference isn't significant. If we change the second coin to $r = .75$, the ratio jumps to 30. If we change it to $r = .8$, it jumps to 4700. These are much more convincing results.

Extending this process to answer our opening question, keep the first coin ($r = .5$) just as it is. Replace the second coin with the average probability value

---

1 Some facts are available with certainty. For example, when we flip a coin of unknown weighting and get a Tail, we can conclude with certainty that this is not an HH coin.

Table 12.5 Interpretation summary of Bayes ratio.

Ratio	Interpretation
1–3	Barely worth mention
3–10	Substantial
10–15	Strong
15–20	Very strong
>20	Decisive

due to all possible values of r. We could approximate this by taking – say – 1000 uniformly distributed values of r, or we could write out the continuous distribution case by replacing the sum with an integral. Since we're probably going to approximate the integral with a finite sum, the result will be essentially the same regardless of which way we go.

Assuming a uniform prior, we find that the ratio of the $r = .5$ case to the average case is 1.1. Interpreting this result, it is probable (but hardly probable) that the coin in question is fair.

There are several interpretations of the "strength" of the ratio available. Table 12.5 was published by Jeffreys.[2]

Now, suppose we have a coin minting machine that produces coins with an average r of .5 but with some random spread in values. We'll characterize the machine's output as Gaussian with a mean of .5 and a sigma that we'll set to several different values so as to see the results. Our question now is whether or not we believe that the coin we are flipping was produced by this machine.

This normal distribution will be the prior distribution for our calculations. From a random sample of the chosen normal distribution (r values), we get a distribution of binomial probabilities of each value of r having produced the chosen value of heads out of the total number of flips. The average of these probabilities is now the numerator in our ratio. The denominator is calculated as in the previous example.

Table 12.6 shows several results. When a ratio is <1, 1/ratio is also shown.

These results satisfy common sense; the smaller the σ of the minting machine, the more tightly grouped the machine's output will be about $r = .5$, and therefore the less likely one of its coins would yield 60 heads out of 100 flips. Also, for a given σ it is harder to be deemed "minted by the machine" if a coin yields 65 heads than if it yields 60 heads (out of 100 flips) and harder still if it yields 70 heads.

---

2 Jeffreys, H. (1961). *The Theory of Probability*, 3e. Oxford.

Table 12.6 Coin minting machine example results.

Nr heads =	60	65	70
σ	Ratio	Ratio	Ratio
.1	2.41	1.49	.74 (1.4)
.05	2.11	.61 (1.6)	.10 (10)
.02	1.32	.15 (6.7)	.007 (142)
.01	1.15	.10 (10)	.002 (500)

Note that, while the calculation of the ratio was rigorous, the interpretation of the results was not. Saying that a ratio of 3 or more is necessary for evidence to be "substantial" is reasonable, but is not derived from basic principles.

Chapter 16 is principally about frequentist statistical inference. However, at the end of Chapter 16, there is a short section comparing Bayesian and frequentist decision making.

The Lighthouse Problem

The lighthouse problem and its mathematical equivalent – the paintball problem – are "standard" examples of parameter estimation using Bayes Theorem. The lighthouse problem is based on a straight coastline with an offshore lighthouse. In our first description, we will assume that the distance from the lighthouse to the (nearest point on the) shore is known.

The light in the lighthouse is mounted in a rotating opaque turret that has a small hole in its side. If the light is on all the time, a narrow beam of light will repeatedly sweep across the shore, as shown (top view) in Figure 12.3.

We will assume that the turret rotates at a fixed rotational speed, i.e. it's connected to a constant speed motor that spins it. However, due to the geometry, the light beam does not sweep along the shoreline at a constant speed. It moves more slowly when the beam is normal to the shore than it does when the beam is almost parallel to the shore.

Our problem derives from this geometry. We know how far the lighthouse is from the shore (at its closest, or normal point) but we do not know where – along the shore – it is located. The beam is not on constantly but rather emits very short bursts of light at random intervals. Along the shore are many photodetectors which record being stuck by light but not the angle that the light came from. In other words, our data set a list of locations (x values) where light strikes. We want to predict the (x value of the) location of the light.

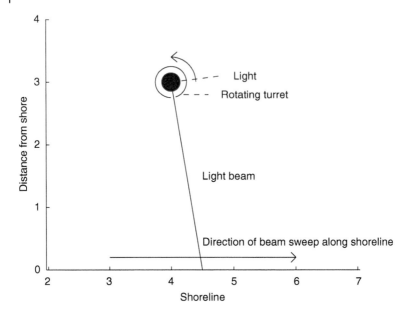

Figure 12.3 Diagram of rotating turret lighthouse problem.

Our likelihood function is the probability of a detector at x_k capturing a flash of light if the light is at x_p. Since – as discussed above – the light spends more time passing a region where its sweeping velocity is lower that it does where this velocity is higher, the probability of a flash being captured by a detector peaks at $x = x_k$.

Without showing the derivation, this probability is given by

$$p(x_k \mid x, y) = \frac{y}{\pi \left[y^2 + (x_k - x)^2 \right]} \tag{12.4}$$

where (x_p, y_p) is the location of the lighthouse and we know y_p.

This function is shown in Figure 12.4.

As a test, a list of uniform random numbers (uniform in the rotation angle of the turret) was generated. The allowed values were from 14° below horizontal facing to the left (Figure 12.3) to 14° above horizontal facing to the right. Since the lighthouse is 1 mile from the shore, this limited the x values of the recorded light flash from (about) 4 miles to the left of the lighthouse to 4 miles to the right of the lighthouse. The lighthouse was located at $x = 3$, but the test "doesn't know this"; our goal is to use the data to find out where the lighthouse is located.

The first few values of x location of the recorded flashes are 2.71, –.38, 2.17, 1.57, and 1.97. Figure 12.5 shows the results of the calculation. The curves are

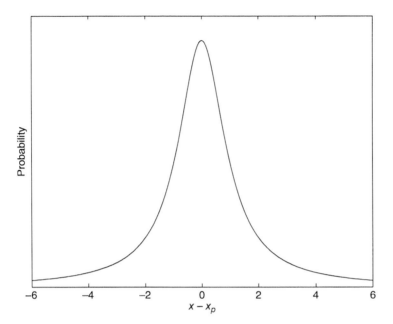

Figure 12.4 The Cauchy distribution.

the posteriors for the number of data points used (from the recorded flashes list), the vertical lines show the locations of these data points.

Referring to Figure 12.5 – when only the first data point is used – the posterior simply reflects "this is our best guess." Using two data points, the results are not so clear or easy to interpret. Using three data points, the posterior curve is beginning to look normal and is narrowing. Then, as the number of data points continues to increase, the curve "settles in" to peaking at the actual location point of the lighthouse and the confidence interval narrowing as the number of data points increases.

The Likelihood Function

The following section contains an interesting calculation but is intended only for readers interested in a bit "deeper dive" into the analysis of this example.

In this section, we will derive the probability Eq. (12.4) and discuss its unusual properties.

Figure 12.6 is essentially a repeat of Figure 12.3, redrawn so that we can see the trigonometric relationships involved more clearly.

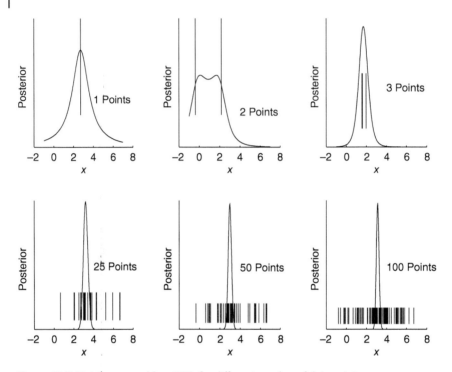

Figure 12.5 Lighthouse problem PDFs for different number of data points.

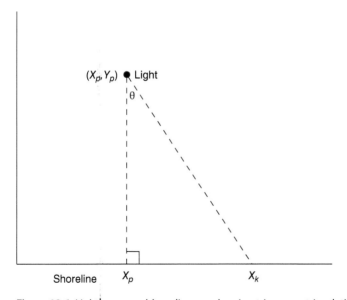

Figure 12.6 Lighthouse problem diagram showing trigonometric relationships.

As the turret spins, the angle θ repeatedly goes from 0 to 2π at a uniform rate. When the light flashes, the position x_k on the seashore is illuminated. From the geometry of the system,

$$\frac{x_k - x_p}{y_p} = \tan(\theta) \tag{12.5}$$

The light only shines on the seashore half the time (the other half of the time it's pointing out to sea). Since the light is rotating at a uniform rate, the probability of it being at a given angle (value of θ) is

$$p(\theta \mid x_p, y_p) = \frac{1}{\pi} \tag{12.6}$$

For the Bayes Formula likelihood, we need $p(x_k \mid x_p, y_p)$.

For two related variables – say u and v – the areas under the probability curves for small changes in u and v must be equal;

$$prob(X \mid I) \, \delta x = prob(Y \mid I) \, \delta Y \tag{12.7}$$

And in the limit as $\delta x, \delta Y > 0$,

$$prob(X \mid I) = prob(Y \mid I) \left| \frac{dY}{dX} \right| \tag{12.8}$$

which leads directly to the relationship we need,

$$prob(x_k \mid x_p, y_p) = prob(\theta \mid x_k, y_k) \left| \frac{d\theta}{dx_k} \right| = \frac{1}{\pi} \left| \frac{d\theta}{dx_k} \right| \tag{12.9}$$

The required derivative is found from Eq. (12.5) and then algebraic/trigonometric manipulation lead directly to Eq. (12.4).

Equation (12.4) is known as the Cauchy distribution. It is clearly symmetric about $x = x_k$. For most similarly symmetric functions, the average of a list of randomly chosen points would converge to the EV. If this were true for the Cauchy distribution, we wouldn't need the Bayesian analysis for the lighthouse problem; just averaging a large number of light flash positions on the shore would lead us to the x position of the lighthouse.

However, the Cauchy distribution is unusual in that it has no EV and no standard deviation! Writing the equations for the formal definitions of these parameters leads to unsolvable expressions. An implication of this is that an average over many points may not converge to the EV as the number of points is increased. Sadly, there is no shortcut to solving the lighthouse problem, you have to calculate the posteriors.

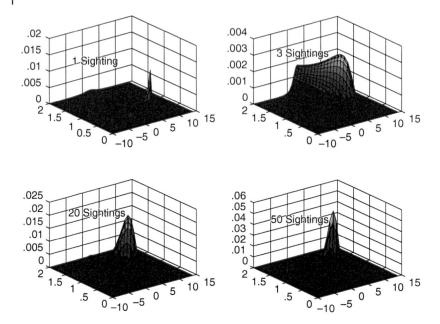

Figure 12.7 3-Dimensional lighthouse problem PDFs for different number of data points.

The Lighthouse Problem II

If neither x_p nor y_p are known, we can still use our probability function, Eq. (12.4) as the likelihood function. However, now we have a function of x and y and must use a grid over which to plot the probabilities. The resulting three-dimensional graphs are shown in Figure 12.7.

Referring to Figure 12.7 – in this example – after one sighting we have a probability peak that tells us more about the sighting than about the lighthouse. After three sightings (remember, this is random angle dependent) we have an ill-defined peak. After 20 sightings we have a definite peak with a reasonable prediction of the lighthouse's location. After 50 sightings the peak is well defined and is narrowing about the correct location ($x_p = 2$, $y_p = 1$).

13

Two Paradoxes

Introduction

A paradox, as defined in dictionaries, is an "apparently false statement that proves to be true." This chapter will present two such paradoxes: Parrondo's paradox and Simpson's paradox.

Parrondo's Paradox

In Chapter 8, we showed that an important factor in gambling games is the expected value of return. If the expected value isn't in your favor, you don't have to look at other factors; in the long run you should expect to lose money.

Juan Parrondo (a physicist), in 1966 described some games of chance in which properly combining two losing games produces a winning game. The first of these games is a simple board game where a player moves back and forth in one dimension, the direction of the move determined by the throw of a pair of dice. Figure 13.1 shows the playing board for this game and Table 13.1 shows two sets of rules for playing the game.

We start using Rule Set 1, shown in the upper-left section of Table 13.1. The player starts on the center (black) square and rolls the dice. The section of Rule Set 1 labeled *black* is used. If the dice roll 11, the player moves forward one box; if the dice roll 2, 4, or 12, the player moves back one box. If the dice roll any other number, there is no movement and the dice are rolled again.

Assume that the player has rolled one of the numbers resulting in a move. The player is now on a white square. The section of Rule Set 1 labeled *white* is used. If the dice roll 7 or 11, the player moves forward; if they roll 2, 3, or 12, the player moves back.

Probably Not: Future Prediction Using Probability and Statistical Inference,
Second Edition. Lawrence N. Dworsky.
© 2019 John Wiley & Sons, Inc. Published 2019 by John Wiley & Sons, Inc.
Companion website: www.wiley.com/go/probablynot2e

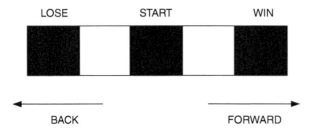

LOSE   START   WIN

BACK   FORWARD

Figure 13.1 Playing board for Parrondo's paradox board game.

Table 13.1 Rule sets for Parrondo's Paradox board game.

	Rule set 1		Rule set 2	
	White	Black	White	Black
Playing rules				
Forward	7, 11	11	11	7, 11
Back	2, 3, 12	2, 4, 12	2, 4, 12	2, 3, 12
Probabilities of plays (36ths)				
Forward	8	2	2	8
Back	4	5	5	4

If the player moves forward to the box labeled WIN then the player has won and the game is over. If the player moves back to the box labeled LOSE then the player has lost and the game is over.

The lower section of Table 13.1 shows the probabilities, multiplied by 36, associated with the various moves shown in the upper section. Since we have left out the (36) denominators, these numbers are actually the number of ways to move, directly proportional to the probabilities of these moves.

Looking at the lower left table section, there are $(8)(2) = 16$ ways to move forward and $(5, 4) = 20$ ways to move backward. The probability of winning is

$$\frac{16}{16+20} = \frac{16}{36} = .444$$

This is definitely not a good game to play.

Rule Set 2 is a bit different from Rule Set 1, but the win-or-lose move counts are exactly the same: 16 to win, 20 to lose, respectively. This is also not a good game to play.

So far there isn't anything remarkable. The rule sets describe two somewhat convoluted but basically uninteresting games, both of which are losing games.

Now let's change things a bit. We'll play the game by starting with either rule set and then randomly decide whether or not to switch to the other rule set after each roll of the dice.

The average number of forward moves is now

$$\left(\frac{8+2}{2}\right)\left(\frac{8+2}{2}\right) = 25$$

and the average number of back moves is

$$\left(\frac{4+5}{2}\right)\left(\frac{4+5}{2}\right) = 20.25$$

The probability of winning is

$$\frac{25}{25+20.25} = \frac{25}{45.25} = .552$$

This is a winning game!

Another approach is to randomly switch rule sets between games according to a random choice based on an assigned probability. Figure 13.2 shows the

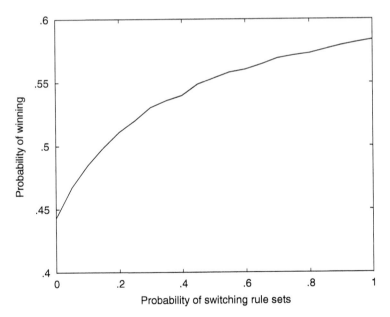

Figure 13.2 Probability of winning Parrondo's paradox board game versus rule set switching probability.

probability of winning this game versus the probability of switching rule sets before each roll.

The winning probability of .552 predicted above occurs for $p = .5$. This is equivalent to flipping a coin before each dice roll to determine whether or not to change the rule set. $P = 0$ is equivalent to never switching rule sets, so the probability of winning is .444 as predicted above, whichever rule set you happen to be using. For $p = 1$, you are switching rule sets before every roll of the dice, and the winning probability climbs all the way to about .58. Note that you only have to switch more than about 20% of the time in order to have a winning game.

Another Parrondo Game

Parrondo's Paradox games do not have to involve probabilities. We can "set up" an example with fixed rules:

Define two games. Game A is very simple: every time you play you lose $1. Game B is almost as simple: every time you play you count your money. If it's an even number you win $3, if it's an odd number, you lose $5.

Table 13.2 shows game A, two variations of game B, and two variations of compound games.

The first game, game A, is trivially simple. It doesn't matter how much you start with, you lose $1 every play.

The second and third games are both game B. If you start with an even amount of money, you gain $3, then lose $5, then gain $3, … . If you start with an odd amount of money, you lose $5, then gain $3, … . In both situations you lose, in the long term average, $1 every play, the same as in game A.

In the fourth game, you alternate ABABA…, starting with an even amount of money. You lose $1 (game A), then lose $5 (game B odd), … . In the long-term average, you lose $3 every play.

Table 13.2 Example plays in Parrondo's Paradox deterministic game.

		Start even	Start odd	Start even	Start odd
Play Nr	Game A	Game B	Game B	Game ABAB	Game ABAB
Start	100	100	101	100	101
1	99	103	96	99	100
2	98	98	99	94	103
3	97	101	94	93	102
4	96	96	97	88	105

In the fifth game, you again alternate ABABA..., but starting with an odd amount of money. You first lose $1 (game A), then gain $3 (game B even), ... In the long-term average, you gain $1 every play.

It's easy to see what is happening here. While choosing game A, every other play costs you $1, it "forces" the choice of winning or losing in game B and can lock you into winning every time game B is played.

There is an infinite combination of games that have not been shown, including BABA starting odd or even, ABBABB starting odd or even, etc.

In all possible cases these are not gambling games, just combinations of algorithms that while individually cause you to lose money, in the correct combinations will cause you to gain money. It is an example of Parrondo's Paradox because we switch between two games (in this case algorithms) and the rules for switching and/or the games themselves requires looking at your history.

Next, consider a Parrondo game set that is a gambling game.

We have two coin flip games, each with their own rule sets. In both games we have the usual bet – the loser gives a dollar to the winner. The first game is very simple, it's just like the typical coin flip game except that the coin isn't a fair coin. The probability of a head is

$$p_1 = .495 \tag{13.1}$$

This is "almost" a fair game. If p_1 is the probability of getting a head and you're betting on heads, you shouldn't plan to win in the long term.

The second game is a bit more complex. There are two more unfair coins. These coins have probabilities of getting a head of

$$p_2 = .095 \tag{13.2}$$

$$p_3 = .745 \tag{13.3}$$

The choice of whether to use the coin with p_2 or p_3 is determined by how much money you've won or lost (your capital, C) when you're about to play the second game. If C is divisible by some integer that was chosen before we started playing (call it M), then use p_2, otherwise use p_3.

We choose which game to play by a random selection with probability P.

We'll use $M = 3$ for our examples. This second game is a bit messier to analyze than the first game, but it can be shown that it's also a losing game.

So far we have two coin flip games that are only good for losing money. But what if we choose which game to play next randomly? Figure 13.3 shows the expected winnings after 100 games versus the probability (P) of picking game one. If $P = 0$ then you're always playing game one and will lose money. If $P = 1$ then you're always playing game two and will always lose money. For a large range of P around $P = .5$ (randomly choosing either game), you can expect to win money. Figure 13.3 is based upon averaging 10 000 iterations of a 100 coin flip game. The "wobble" in the curve points indicates a very large standard

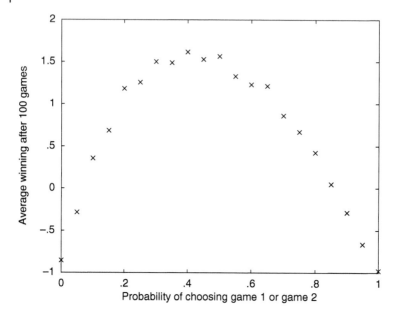

Figure 13.3 Probability of winning Parrondo's paradox gambling game(s) versus probability of game choice.

deviation in these (Monte Carlo) simulations, even after averaging over 10 000 sets of games.

Why does this work? Game 2 is structured so that more times than not you're losing. However, whenever you get to a certain level of capital, you win. Game 1 is losing, but barely losing. Jumping from game 2 to game 1 after winning game 2 gives you a reasonable probability of "locking in" this gain. If the probabilities are structured correctly, you win game 2 often enough and lock in the winnings often enough, and you come out net ahead.

Can you go to Las Vegas or Atlantic City and set yourself up with some set of games that are equivalent to the above? It doesn't appear so because of the dependency of the odds of game 2 on previous results, which sometimes come from game 1. Also, you wouldn't be able to piece this together yourself from several different games because one of the games has a probability of winning of close to 75%. If you ever find such a game, don't worry about Parrondo's Paradox – buy a home near the casino that's offering this game and make playing it your life's work.

On the other hand, suppose you did talk some casino owner into setting you up with this combination of games. Would they be really stupid in giving you (a set of) games that have the odds in your favor?

The standard deviations on each of the results is very large as compared to the expected values; the results will be quite erratic. Although the odds are truly in your favor, they're only very slightly in your favor. This is beginning to look like the "unlosable" game of Chapter 8: the mathematics says that you won't lose, but you have to start out with an awful lot of money in your pocket to cover the variations if you hope to realize the game's potential.

One last thought on this dice game. We can find a game in a casino where the winning probability is almost .5 – for example, betting on one color at a roulette table. We can certainly find a game where the winning probability is almost .1; if we can't find one we could certainly get a casino owner to create such a game for us. What about completing the picture by having a friend create the game where the winning probability is almost .75? We would fund this friend out of the winnings, and now we're set to run the complete Parrondo dice game.

Unfortunately, this would be a total disaster. What happens is that we win a little bit, the owner of the "almost .5 probability game" wins a little less than us, the owner of the "almost .1 probability game" wins about 20 times more than us, and our friend, the surrogate owner of the "almost .75 probability game" funds it all. Oh, well.

The Parrondo Ratchet

The "Parrondo Ratchet" is another example of two "losing" situations that combine to create a "winning" situation.

If some balls are placed close together on a level surface that is being shaken back and forth slightly, they will behave according what we have described as Random Walks, AKA Brownian Motion. There will be an equal average of motion in all directions. If the surface is tilted upward on, say, the right side, then balls moving to the left will pick up speed (on the average) and balls moving to the right will lose speed (on the average). All in all, there will be a net motion to the left, i.e. down the hill. Note that we must always stick in the term On The Average because the shaking of the surface combined with the collisions of the balls can sometimes reverse the direction of travel of a ball for a short time.

What if we build a sawtooth structure, also known as a linear ratchet, as shown in Figure 13.4, shake it a bit, and drop some balls onto it? If a ball has enough energy to climb over the top of the local hill, then it would move laterally from hill to valley to hill to valley, etc. If we have a bunch of balls rattling around with enough total energy for a ball to occasionally climb over the top of a local hill, then we would see a gradual diffusion of the ball collection (much like the ants in Chapter 3). If the ratchet is tilted so that it is lower on the right than on the left, we would expect to see a net motion of balls to the right, i.e. "down the hill."

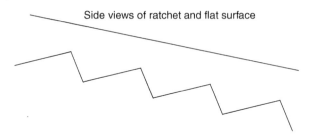

Side views of ratchet and flat surface

Figure 13.4 Sketch of sloped ratchet and ramp for Parrondo's paradox "Free Energy" discussion.

Now, let's assume that the tilted plane and the tilted ratchet are side by side and balls are getting knocked back and forth between the two as well as moving up and down the slope. Given the right parameters, this situation closely mimics the Parrondo coin toss game. The tilted plane is game 1, the ratchet, with two different sets of rules (the asymmetric teeth), is game 2. And, just as in the Parrondo coin toss, things can go contrary to what you'd expect: there can be net movement of balls up the hill.

The Parrondo ratchet system is the subject of study of people trying to learn how and if it is possible to extract energy out of a system by harnessing random motion. Clearly if they're looking at Brownian motion, they're going to need some very small ratchets, but this is the age of nanomechanics. The problem here is that the second law of thermodynamics tells us that we cannot extract work from a thermal machine which is at a unit temperature throughout. If your analysis of your ratchet and gears and wheels and …, tells you that you can, then your analysis of your system is somehow in error.

Simpson's Paradox

Simpson's paradox refers to an unintuitive result that comes out of certain simple algebraic maneuvers. Simpson's paradox occurs in many common situations whereas Parrondo's paradox, either the probability game or the coin flip version, is more commonly restricted to academic studies. We'll introduce the topic with an example; this example is based on a real situation at U.C. Berkeley some years ago.

A school is introducing two new programs this year. We'll call these programs I and II. Each program has room for 100 students. If more than this number apply, the admissions director must choose the best qualified applicants.

Program I is very popular; 600 women and 400 men apply (for admission) to program I. Remember, there are openings for 100 students in program I.

75 women and 25 men are accepted into the program. The rates of acceptance are 75/600 = 12.5% of the women applicants and 25/400 = 6.25% of the

men applicants. The rate of acceptance of women is double that of the rate of acceptance of men.

25 women and 75 men apply to program II. Program II also has 100 openings, so all of the applicants are accepted. The rate of acceptance of women equals the rate of acceptance of men = 100%.

Adding up the numbers for the two programs, 625 women applied and 100 women were accepted. The total rate of acceptance for women was 100/625 = 16%. 475 men applied and 75 men were accepted. The total rate of acceptance for men was 75/475 = 21%.

Even though the rates of acceptance for women were the same or better than the rates of acceptance for men in both programs, the overall rate of acceptance for men was better than the overall rate of acceptance for women. In situations such as this, there is a *lurking variable* that gets lost in the summarization process. In this particular example, the lurking variable is the tremendous disparity in the number of applicants to the two programs.

Before looking at another example, let's formalize the arithmetic. Call the two initial categories of data, the men and the women, small abcd and capital ABCD, respectively. Let

a = the number of men accepted to program I (25)
b = the number of men that applied to program I (400)
c = the number of men accepted to program II (75)
d = the number of men that applied to program II (75)
A = the number of women accepted to program I (75)
B = the number of women that applied to program I (600)
C = the number of women accepted to program II (25)
D = the number of women that applied to program II (25)

In this notation, the statement that the percentage of men accepted to program I is less than the percentage of women accepted to program I is expressed as

$$\frac{a}{b} < \frac{A}{B} \tag{13.4}$$

The statement that the percentage of men accepted to program II is the same as the percentage of women accepted to program II is expressed as

$$\frac{c}{d} = \frac{C}{D} \tag{13.5}$$

And the statement that the percentage of men admitted to both programs is greater than the percentage of women admitted to both programs is:

$$\frac{a+c}{b+d} > \frac{A+C}{B+D} \tag{13.6}$$

Let's consider another example. In medical literature, the term *confounding variable* is used rather than the *lurking variable* seen in probability and statistics literature. In the following example, the confounding variable is the size of kidney stones.

We have two treatments for kidney stones, I and II. For small stones, out of 87 patients treated with treatment I, 81 results were considered successful. Using the notation introduced above,

$$\frac{A}{B} = \frac{81}{87} = 93\% \tag{13.7}$$

270 patients were treated with treatment II and 234 of these were considered successful,

$$\frac{a}{b} = \frac{234}{270} = 87\% \tag{13.8}$$

We can conclude that treatment I is the more effective of the two for small stones since

$$\frac{a}{b} < \frac{A}{B} \tag{13.9}$$

For large stones, 263 patients received treatment I with 192 successes,

$$\frac{C}{D} = \frac{192}{263} = 73\% \tag{13.10}$$

and 80 patients were treated with treatment II, yielding 55 successes.

$$\frac{c}{d} = \frac{55}{80} = 69\% \tag{13.11}$$

Again, treatment I is the more effective treatment:

$$\frac{c}{d} < \frac{C}{D} \tag{13.12}$$

We seem to have a clear winner in the choice of treatments. Treatment I is more effective than treatment II for both large and small stones. However, if we try to summarize this conclusion by adding up the numbers,

$A + C = 81 + 192 = 273 =$ treatment I successes out of $B + D = 87 + 263 = 350$ attempts. Therefore, for treatment I overall, the success rate is

$$\frac{A+C}{B+D} = \frac{273}{350} = 78\% \tag{13.13}$$

For treatment II, we have $a + c = 234 + 55 = 289$ successes out of $b + d = 270 + 80 = 350$ attempts. Therefore, for treatment II overall, the success rate is

$$\frac{a+c}{b+d} = \frac{289}{350} = 83\% \tag{13.14}$$

Again, we find that

$$\frac{a+c}{b+d} > \frac{A+C}{B+D} \tag{13.15}$$

Based on the summary it is "clear" that treatment II is more effective.

The natural question to ask now is "what went wrong?" From a strict mathematical point of view, nothing went wrong. There are no errors in the calculations above. Why, then, in each example, did we (seem to) get two different answers? Because we asked two different questions. In both examples, the first question asked, dealing with the two sets of numbers separately, is what it seems to be. The second question asked, dealing with the two sets combined, is not what most people want it to be (or think it is).

Leaving this point as a cliff-hanger for the moment, let's go back to the numbers.

Table 13.3 repeats the definitions and numbers for the college programs admission example. The column "Orig Nrs" is a repeat of the data and calculations presented above.

Table 13.3 Simpson's Paradox school admission example.

Item	Explanation	Orig Nrs	Modif. Nrs
a	Nr men accepted to program I	25	25
b	Nr men applied to program I	400	400
c	Nr men accepted to program II	75	75
d	Nr men applied to program II	75	750
A	Nr women accepted to program I	75	75
B	Nr women applied to program I	600	600
C	Nr women accepted to program II	25	25
D	Nr women applied to program II	25	250
	Calculations		
	a/b	.06	.06
	c/d	1.00	.10
	A/B	.13	.13
	C/D	1.00	.10
Concl I	$[(a+c)/(b+d)]/[(A+C)/(B+D)]$	1.32	.74
Concl II	$[a/b + c/d]/[A/B + C/D]$	.94	.71

The column "Modif. Nrs" is the same as the column to its left except that we've increased the applicants to program II (d and D) by a factor of ten. The acceptances have not changed.

The row labeled Concl(usion) I is the summary as described above, with the Orig Nrs and the Modified Nrs shown. As presented above, the Orig Nrs is 1.32, indicating that men are preferred for admission over women. However, the same calculation for Modified Nrs shows that women are preferred for admission over men.

In other words, the summary is not logically consistent with the conclusions for the separate programs; it isn't even consistent with itself as the size of (one of) the groups is changed. Note that in Modified Numbers the number of applicants to the two programs are comparable, whereas in Original Numbers they are very different. This can be mapped out in greater detail, the conclusion being that Simpson's paradox comes about when these big differences exist. Again, it's not an error; it's just that that the summary calculation isn't saying what you want it to say and the result of the summary calculation, while arithmetically correct, doesn't mean what you want it to mean.

The bottom row of Table 13.3, Concl(usion II) is another way of looking at things. a/b is the ratio of the number of men accepted to program I to the number of men applying to program I; it's the probability of a man who applies to program I being accepted to program I. Similarly, c/d, A/B, and C/D are acceptance probabilities. The calculation of the Conclusion II row is the ratio of the average of the probabilities for men to the average of these probabilities for women. As the last two columns shows, these numbers will change as group sizes are changed, but they will not "cross through" one. The conclusion is that (the result of) this calculation is a function of group sizes but is always less than one; women applicants are preferred over men applicants, consistent with the sub-category conclusion.

Problems

This chapter has a "minimal" set of problems. There are two reasons for this:

1) Parrondo's paradox is considered an interesting topic in that it violates the guideline about not gambling in a game where the Expected Value of your profits is negative. We could certainly put together other examples and couch them as problems, but this is a curiosity that almost nobody will ever encounter "out in the real world."
2) Simpson's paradox is a real world issue. The examples in the text and in Problem 13.1 are taken from real situations. The nature of this "paradox" is however very clear, and the two included problems demonstrate it. We could certainly generate more problems, but they would simply be a rehash of Problems 13.1 and 13.2.

13.1   This is real historical data.

Information probably needed by nonbaseball fans: Each season, a player will be a batter ("come to bat") some number of times ("nr of at-bats"). He will successfully hit the ball some fraction of this nr of at-bats, his "nr of hits." The ratio of these is his "batting average" which is considered a strong measure of how well this player performed in a season.

In 1995, Derek Jeter came to bat 48 times and got 12 hits. David Justice came to bat 411 times and got 104 hits. In 1996, Derek Jeter came to bat 183 times and got 582 hits. David Justice came to bat 45 times and got 140 hits.

A For each year, which batter had the higher batting average?

B For the two years combined, which batter had the higher batting average?

13.2   Referring to Problem 13.1, give an example of sets of numbers that would keep both player's individual year performances the same but would also show Justice's combined year performance as superior.

14

Benford's Law

Introduction

Benford's law refers to some fascinating statistics that occur in the natural world, i.e. they were discovered, not created. These statistics are fundamentally different than anything discussed thus far in that they refer to individual digits of numbers (principally the first digit), not to the numbers themselves. For example, 123.2, 1.45, and 1 500 000 all have the same first digit, whereas 223.2 has a different first digit than the first three numbers. The occurrences and properties we will learn about from looking at these numbers are unlike anything examined (in this book) before.

History

Before the advent of electronic computers and pocket calculators, most mathematics/physics/engineering calculations were performed using logarithms. The simplest way to explain logarithms without actually deriving the mathematics is to say that logarithms are a system which reduces multiplication and division to addition and subtraction and raising a number to a power or to a root reduces to multiplication or division. This latter attribute implies that it is only necessary to know the logarithms of numbers between one and ten. The logarithm of thirty, for example, is just the logarithm of ten times three, which in turn is the logarithm of ten plus the logarithm of three.

Logarithms of the numbers from one to ten, to many decimal places, were published in weighty books. A book of these tables of logarithms was found on the desk of just about every working engineer, physicist, and applied mathematician.

Late in the nineteenth century, Simon Newcomb, an American astronomer, made a curious observation. Looking at his well-worn table of

Probably Not: Future Prediction Using Probability and Statistical Inference,
Second Edition. Lawrence N. Dworsky.
© 2019 John Wiley & Sons, Inc. Published 2019 by John Wiley & Sons, Inc.
Companion website: www.wiley.com/go/probablynot2e

logarithms book, he noticed that the pages at the front of the book were much more worn than were the pages at the back of the book. In other words, over the years he had looked up the logarithms of numbers beginning with the digit 1 more than he had looked up the logarithms of numbers beginning with the digit 9. And, the apparent usage dropped monotonically through all the first (leading) digits throughout the book – i.e. from 1 to 9. In the book itself there are, of course, equal numbers of pages beginning with each of the nine digits.

Newcomb learned that this was not just an anomaly of his book. It was the common situation in well-worn table of logarithms books.

Newcomb proposed that this relationship comes about due to the occurrence of numbers with these leading digit statistics in nature. He proposed that the relative occurrence of leading digits obeys

$$P(n) = \log\left(1 + \frac{1}{n}\right) \tag{14.1}$$

This relationship is shown in Figure 14.1.

In 1938, this relationship was again reported by Frank Benford, a physicist. As sometimes happens in these situations, the relationship, Eq. (14.1), is usually attributed to Benford rather than Newcomb. Benford, to be fair, also gathered a large amount of data and demonstrated other important properties of the distribution.

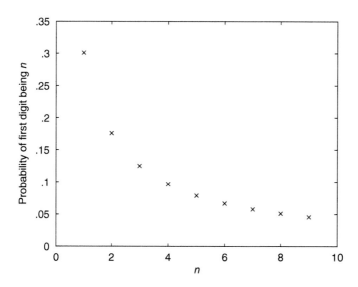

Figure 14.1 Benford first digit coefficients.

The 1/x Distribution

Before presenting and discussing these properties, let's step back and consider a probability distribution that obeys Eq. (14.1).

Figure 14.2 shows two functions: $f_1(x) = \frac{1}{x}$ and $f_2(x) = \frac{1}{x^2}$, for $1 \le x \le 10$. While these are different functions, they are similar enough to allow some important comparisons. Note that these functions have not been normalized, so they are not proper PDFs.

Table 14.1 shows the leading digit statistics from large lists (100 000) of numbers generated using the above functions as distributions along with the values predicted by Eq. (14.1).

The $\frac{1}{x}$ function numbers look very much like the equation numbers. The $\frac{1}{x^2}$ numbers, on the other hand, do not. Since both functions are monotonically decreasing functions of x, it is not surprising that the leading digit occurrences fall as the leading digit increases.

Now we'll do something that anticipates one of the most fascinating properties of Benford statistics; we'll take all the numbers that went into the data in the above table and double them. Since the original data sets had contained numbers between 1 and 10, the new sets contain numbers between 2 and 20. Next we'll repeat collecting the leading digit statistics and present them in Table 14.2.

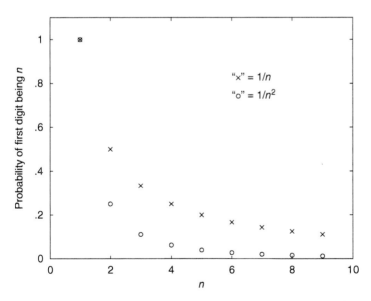

Figure 14.2 Comparison of leading digit statistics For Benford's series, $\frac{1}{n}$ and $\frac{1}{n^2}$ series.

Table 14.1 Comparison of leading digit statistics for Benford's series, $\frac{1}{n}$ and $\frac{1}{n^2}$ series.

Digit	Equation (14.1)	1/n	1/n²
1	.3010	.301	.555
2	.1761	.178	.185
3	.1249	.124	.093
4	.0969	.095	.056
5	.0792	.079	.037
6	.0669	.067	.027
7	.0580	.058	.020
8	.0512	.051	.015
9	.0458	.046	.012

Table 14.2 Comparison of leading digit statistics for Benford's series, $\frac{1}{n}$ and $\frac{1}{n^2}$ series after scaling.

Digit	Equation (14.1)	1/n	1/n²
1	.3010	.302	.111
2	.1761	.176	.371
3	.1249	.125	.184
4	.0969	.098	.110
5	.0792	.080	.074
6	.0669	.067	.055
7	.0580	.058	.020
8	.0512	.051	.031
9	.0458	.046	.025

This table shows that the leading digit statistics of the $\frac{1}{x}$ function is still pretty much the same as it was. The leading digit statistics of the $\frac{1}{x^2}$ distribution, on the other hand, is wandering around aimlessly.

Questions that immediately arise are: is what just happened a coincidence? What if we had multiplied by some number other than 2? What if the data sets had been for some range of numbers other than from 1 to 10?

In order to address these questions, we must examine some properties of the $\frac{1}{x}$ function.

Note: If you are uncomfortable with the mathematics of this analysis, just skip through it. The conclusions are easy to understand; you'll just have to accept that they were properly derived.

The fraction of a distribution of leading digits based on uniformly distributed random variables in the range $a \le x \le b$ is found by calculating the number of variables with this leading digit and dividing this number by the total number of variables expected from the same set. For the leading digit n, where n is an integer, $1 \le n \le 9$, and the $\frac{1}{x}$ distribution, this fraction is

$$B_n = \frac{\int_n^{n+1} \frac{dx}{x}}{\int_a^b \frac{dx}{x}} \tag{14.2}$$

As an example from Table 14.1, let $a = 1$, $b = 10$, $n = 1$:

$$B_1 = \frac{\int_1^2 \frac{dx}{x}}{\int_1^{10} \frac{dx}{x}} = \frac{\ln(2) - \ln(1)}{\ln(10) - \ln(1)} = .3010 \tag{14.3}$$

For all values of n, this becomes Eq. (14.1).

In other words, for $a = 1$, $b = 10$, the $\frac{1}{x}$ distribution obeys the Benford's law leading digit statistics.

On any one decade range, for example, $5 \le x \le 50$, we have

$$B_1 = \frac{\int_{10}^{20} \frac{dx}{x}}{\int_5^{50} \frac{dx}{x}} = \frac{\ln(20) - \ln(10)}{\ln(50) - \ln(5)} = \frac{\ln(2)}{\ln(10)} = .3010 \tag{14.4}$$

In other words, the decade range's starting point doesn't matter.

What about a two decade range? For example, $a = 12.5$, $b = 1250$:

$$B_1 = \frac{\int_{12.5}^{20} \frac{dx}{x} + \int_{100}^{200} \frac{dx}{x} + \int_{1000}^{1250} \frac{dx}{x}}{\int_{12.5}^{125} \frac{dx}{x}} = \frac{\ln\left(\frac{20}{12.5}\right) + \ln\left(\frac{200}{100}\right) + \ln\left(\frac{1250}{1000}\right)}{\ln(100)}$$
$$= .3010 \tag{14.5}$$

This can be extended through as many decades as desired, the result does not change.

On the other hand, for a piece of a decade, say, $a = 1$, $b = 5$,

$$B_1 = \frac{\int_1^2 \frac{dx}{x}}{\int_1^5 \frac{dx}{x}} = \frac{\ln(2)}{\ln(5)} = \frac{\ln(2)}{\ln(5)} = .4307 \tag{14.6}$$

The $\frac{1}{x}$ distribution only obeys the Benford's law leading digit statistics for multiples of and combinations of decade ranges.

However, consider the range $a = 1$, $b = 5 \times 10^m$, where m is a positive integer

$$B_1 = \frac{m\int_1^2 \frac{dx}{x}}{\int_1^{5 \times 10^m} \frac{dx}{x}} = \frac{m\ln(2)}{\ln(5) + m\ln(10)} \tag{14.7}$$

As m gets large, this approaches .3010. This can be generalized to $b =$ any large number. In other words, for very large ranges (multiple decades) the exact ends of the range don't matter.

At the other extreme, any distribution which is less than a decade wide can never satisfy Benford's law. Even if we start out with a carefully crafted fake, e.g. $1.5 \le x \le 9.5$, with the shape of the curve properly distorted from the $\frac{1}{x}$ distribution so as to give the correct statistics, there will be some multiplier which leaves the gap in contributions at a crucial interval and the scheme falls apart.

The entire discussion of Benford's law thus far has been simply a mathematical curiosity. However, does it relate to anything in the real world? It seems like it should, based on Newcomb's original observation of the worn pages in his table of logarithms.

Surface Area of Countries of the World

Table 14.3 shows the first few and the last few entries on a list of the countries of the world and their surface areas, in square-kilometers (km^2). There are 260 entries in the full table.

Table 14.4 shows the leading digit statistics for this list of countries. The column labeled Fraction is based directly on data of Table 14.3.

While not an exact match, these numbers look suspiciously like the Benford coefficients. The data of Table 14.3 was based on km^2 data. Multiplying this data by a constant is equivalent to changing the measurement units. For

Table 14.3 Excerpt from countries of the world and their land areas table.

Rank	Country	Area, km^2
1	Russia	17 098 246
2	Antarctica	14 000 000
....		
...		
259	Monaco	2 (rounded)
260	Vatican City	1 (rounded)

Table 14.4 Leading digit statistics from countries of the world and their land areas table.

Nr	Fraction	Fraction from scaled data
1	.285	.324
2	.189	.165
3	.131	.123
4	.104	.108
5	.0692	.0577
6	.0731	.0577
7	.0614	.0538
8	.0385	.0423
9	.0500	.0500

example, there are .3861 square miles in a square kilometer. Therefore, if we multiply every surface area number in Table 14.3 by .3861, we will have converted the data from square kilometers to square miles.

In square miles, the leading digit statistics for the surface area of the countries of the world are shown in the right hand column of Table 14.4 ("Fraction from scaled data"). The statistics have moved around a bit, but they still certainly "look like" the Benford coefficients.

Goodness of Fit Measure

At this point we need a calculation to measure the "goodness of fit" of a set of numbers to the (exact) Benford coefficients. "It looks pretty good to me" does not meet this criterion.

There are many ways that this calculation can be approached. Perhaps the most direct way would be to sum the differences (or squares of the differences) between the leading digit statistics and the exact (Benford) statistics. This will work, but it leaves open a question of weighting: is a 10% (say) error in the ones digit more, less, or equally important than a 10% error in the much more sparsely populated nines digit?

The following calculation is based upon Steven Smith's overall discussion of Benford's law.[1] We'll have a lot more to say about Smith's contribution to our understanding of Benford's law later in the chapter.

---

1 Steven W. Smith (1997). Chapter 34, Explaining Benford's law. In: *Digital Signal Processing*, 1e. California Technical Pub..

We begin by recording the percentage of numbers in our data set whose first digit is 1. Remember that the Benford coefficient for this number is 30.10%.

Next, we multiply every number in our data set by 1.01. Then we repeat the above procedure.

We repeat this 232 times. When we are done, we will have multiplied every number in our data set by $1.01^{232} \approx 10$. We now have a list of 232 numbers representing the fraction of first digits that are equal to 1 at each multiplication. After multiplying a number by ten, its first digit has returned to where it was when we started.

For an ideal fit to the Benford distribution, every one of these numbers would be .3010. The worst possible case would be one or more of these numbers = 1.0. We can therefore define an error for the series by finding the largest number in the list, and calculating

$$\frac{\max(L) - .3010}{.699}, \tag{14.8}$$

where L is the list of numbers. The .699 in the above formula scales the result so that a worst possible case answer would be 1.0.

Equation (14.8) goes to zero for an ideal fit and to one for the worst possible fit. The problem with it is that the numbers get "compressed" for very good fits and it is hard to see small changes in good fits.

We correct this problem by defining our error function as the inverse of the above,

$$OnesErr = \frac{.699}{\max(L) - .3010} \tag{14.9}$$

Figure 14.3 shows these 232 numbers for 1000 points randomly chosen from a $\frac{1}{x}$ distribution on $1 \leq x \leq 100$.

In this example, OnesErr ~ 25.9. This number will of course never repeat exactly because we're studying the output of (pseudo) random number generators.

Repeating the above calculation for the $\frac{1}{x^2}$ distribution we get OnesErr = 2.83, clearly not as good as the result for the $\frac{1}{x}$ distribution.

We now have a tool for estimating how "Benfordish" a real data set is; we can compare it to the results for the $\frac{1}{x}$ distribution, realizing that this is a "soft" rather than an exact metric. Remember that the comparisons must be made for similar sized data sets.

Returning to the countries' surface area data set. Following the same procedure, we get OnesErr of 15.5. Note that the One Scaling Test is actually looking at the invariance of the Benford coefficients for a data set. The resulting OnesErr number therefore does not change if we enter the data set in square kilometers, square miles, or any other set of units.

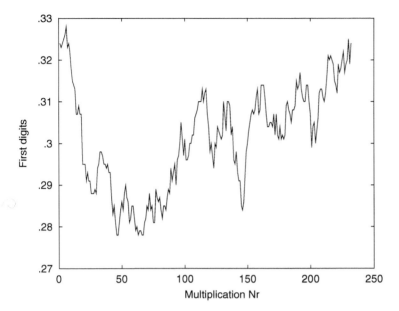

Figure 14.3 Ones error calculation for 232 numbers randomly chosen from $\frac{1}{n}$, $1 \leq x \leq 100$.

The United Nations publishes population numbers for the countries of the world each year. From the 2017 population data, which varies from 1.4 billion in China down to 792 in Vatican city, we get a OnesErr of 12.66. It looks as if something is going on here; data gleaned from the real world seems to "like" Benford statistics.

The discussion as to why many naturally occurring distributions obey Benford's law went on for many years and was not resolved until Steven Smith published a chapter on this topic in the aforementioned reference. Smith's book is about Digital Signal Processing – a topic not often delved into by the statistics community. Smith himself comments that he was looking for some connection of Benford's law to some digital signal processing questions, but failed to find this connection. However, he did explain why Benford's statistics seem to show up in the natural world.

Smith's Analysis

Jumping ahead a bit, Smith's answer as to why many naturally occurring distributions obey Benford's statistics is very simple: "They don't, it just seems that way." This, of course, requires some discussion.

Smith approaches the analysis of Benford statistics as a digital signal processing problem. This is probably unique in the field of statistics. His analysis is not difficult to follow, but it requires familiarity with topics such as Fourier transforms and convolutions integrals. These topics are outside the scope of this book, so we will just quote some important results.

We begin by examining the log-normal distribution. The reason for this choice will be shown soon.

The set of numbers

$$X = e^{\mu+\sigma N} = e^{\mu}e^{\sigma N} \tag{14.10}$$

where N is a normal distribution with mean and sigma μ, σ respectively, will be a log-normal distribution with parameters μ, σ. As the second form of this equation shows, e^{μ} is just a scaling factor, and, when dealing with Benford statistics scaling factors don't alter the validity of the statistics. Therefore, we lose no generality if we set $\mu = 0$ so that $e^{\mu} = 1$.

Figure 14.4 shows several examples of the log-normal distribution, for sigma = .25, .50, and .75. A curve of $1/x$ is also included, for comparison. These curves bear a "family resemblance" in that, starting at $x = 1$, they are all monotonically decreasing functions. Otherwise, there is nothing interesting to comment on, neither resemblances nor differences.

Figure 14.5 were generated using 10 000 point simulations. In Figure 14.5a the Ones Error is plotted versus σ. At $\sigma \approx .7$ the lognormal distribution approximately demonstrates Benford statistics. As σ gets larger, this approximation

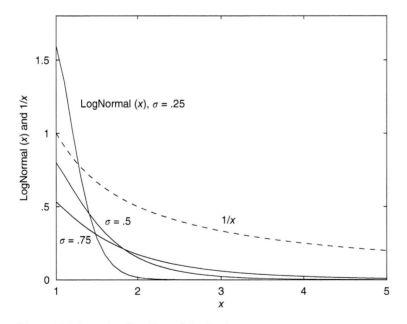

Figure 14.4 Example of LogNormal distributions.

(a)

(b)

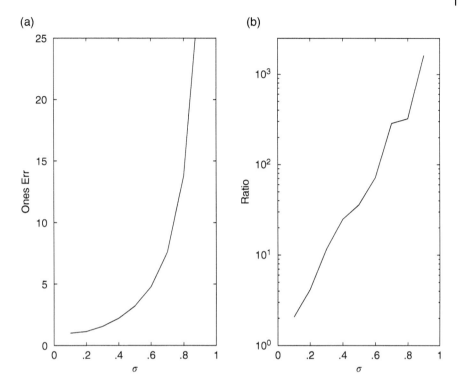

Figure 14.5 OneErr and ratio of largest : smallest number for σ of LogNormal distribution.

gets better, eventually reaching the point where its (the lognormal distribution's) agreement with Benford statistics is indistinguishable from that of the $1/x$ distribution.

Figure 14.5b shows the ratio of the largest number to the smallest number for each simulation point. As σ gets larger, this ratio grows. In the range of σ where the OnesErr is large, this ratio is several decades. This is in agreement with the previous conclusion that a distribution obeying Benford statistics must span at least several decades.

Reviewing the above conclusions, for σ equal to or greater than $\approx.8$, the Benford statistics of a log-normal distribution, for a finite sized data set, are indistinguishable from those same statistics drawn from a similarly sized data set that satisfies Benford's criteria exactly. Saying this slightly differently (for sigma equal to or greater than one and finite size data sets), while the Benford statistics of a log-normal distribution are not exactly those of a Benford data set, they are indistinguishable from those of an exact Benford data set.

Why is this important? In Chapter 3, it was shown that the sum of random data sets approaches a normal distribution. From this it is derivable that the product of random data sets approaches a log-normal distribution.

Table 14.5 Benford statistics from multiplying uniform distributions.

Nr of mults	Range	OnesErr
1	1:100	1.3
2	1:1000	5.3
3	1:10000	25.

As an example, consider uniform distributions of ten integers with values between 1 and 10. If we multiply two of them together, we get a distribution of 100 numbers with a possible distribution between 1 and 100, etc.

Table 14.5 summarizes the results of three such multiplications.

This, then, is our connection between naturally occurring data such as the population of countries and Benford statistics. Just as some naturally occurring data set came about from the sum of other random number data sets, some of them came about from the product of other random number data sets. And, if these latter data sets encompass a wide enough range, they will display first digit statistics that are indistinguishable from Benford statistics.

Just as the sums of arbitrary data sets will approach normal distributions as more sets are added in, the products or arbitrary data sets will approach lognormal distributions. As engineers or physicists or ..., as we go to work with the data sets we find in nature, some of them are the products of other data sets; with these we will find that about 30% of the numbers begin with 1 while only about 5% of them begin with 9. That's a 6 : 1 ratio. No wonder the front of the logarithm book sees much more wear than does the back!

The 2017 U.S. census of population has 3142 entries. These run from the greater than 10.1 million population of Los Angeles Country, California, to the 88 population of Kalawao County, Hawaii; the range of this data set is 115 000 : 1. The OnesErr number for this data set is a very impressive 31.3. For all practical purposes, this is as good as it gets.

If we shrink the size of this data set by a factor of ten by just "pulling out" every tenth point and forming a new data set, the OnesError number only falls to 29.3, still an excellent result.

Now let's try various approaches to corrupting some data and see how it affects the Benford statistics.

First, remember that multiplying every number in the set by a constant (scaling) won't change the OneError at all; the fact that this data set scores so high on the OnesError test in the first place tells us this. But, what about some simple error in collecting data? Suppose each element of the data set is multiplied by a normal random variable generated using $\mu = 1$ and some specified σ? Setting $\mu = 1$ means that, on the average, the data hasn't been perturbed. A non-zero σ means that each number will be perturbed somewhat. The value of σ correlates to how carefully the data was taken.

Table 14.6 Degrading Benford statistics by "Scrambling" data set with normal variation.

σ	OnesErr
.001	29.3
.005	29.3
.01	23.1
.05	20.9
.1	19.1

Table 14.6 shows the result of this calculation. As would be expected, for σ very small, there is no degradation in the OnesErr. Then, as σ grows, degradation appears and gets worse.

A similar degradation can be seen with additive offsets, either regular or random with some prescribed distribution.

A few properties of Benford's law that haven't been mentioned are:

1) Benford's law works regardless of the base of the numbers used. The coefficients are of course different in different bases.
2) Not only are there leading digit statistics, there are second digit, and third digit, and ... statistics. As you go from the leading digit to the second digit to ......, the statistics become progressively more uniform. Knowing all of these properties is important when doing "forensic accounting."
3) Benford's law is finding considerable use in the area of financial fraud detection. For example, the travel expense statements of a large corporation over a year or two will obey Benford's law. A manipulation of the numbers will quickly become apparent when the statistics are analyzed. This makes the IRS, possibly the last group you'd think would be interested in Benford's law statistics, to be one of the leading experts.

Problems

14.1 Equation (14.7) shows that as the first digit = 1 fraction in a $1/x$ distribution will approach .3010, the Benford coefficient, over the range $1 \leq x \leq 5 \times 10^m$ as m, an integer, gets large.

A How large does m have to be for the error in this approximation to be less than 10%?

B We can repeat the above calculation, replacing the 5 by any real number less than 10, and expect similar results. What about the lower limit of the range (=1 in the above example)?

14.2 By extension of Eq. 14.2, derive the second digit Benford coefficients.

15

Networks, Infectious Diseases, and Chain Letters

Introduction

Some diseases propagate by your casual encounter with an infected person, some by your breathing the air that an infected person just sneezed or coughed into, and some by intimate contact with an infected person. In any situation, the statistics of the disease propagation (and possible epidemic) is dependent upon the person–person network; e.g. you will only give AIDs to (some of) the people that you have unprotected intimate contact with. The subject of inter-people networks has been popularized in recent years by the notion of "6 degrees of separation" connecting every person to every other person on the earth. Just what this means and what the choices are in modeling these networks is an interesting subject and is worthy of discussion.

"Stuff" propagating along networks is a multifaceted topic. In addition to disease infection, propagation there is information flow along the internet, electric power flow along the national grid, water flow through canals, etc. On the more whimsical side, there's the chain letter.

Understanding infectious disease propagation can literally be of life and death importance while following the flow of dollars and letters in a chain letter's life is really more a curiosity. However, both of these subjects make interesting studies.

Degrees of Separation

If we want to study how an infectious disease propagates, we need to study the statistics of how people interact. A full detailed model of who you see on a regular basis, how many hours you spend with who, who you touch, who you sneeze and cough at, and then who these people spend time with and sneeze

Probably Not: Future Prediction Using Probability and Statistical Inference,
Second Edition. Lawrence N. Dworsky.
© 2019 John Wiley & Sons, Inc. Published 2019 by John Wiley & Sons, Inc.
Companion website: www.wiley.com/go/probablynot2e

and cough at is an incredibly big job. Here we will try to simplify the situation to its essential components. This will let us see what's happening and understand the issues involved, albeit at the cost of accuracy and some subtleties.

There are many ways to look at people–people interactions. The most well-known of these is the "6 degrees of separation" concept:

Suppose you know 25 people. Let's call this relationship 1 degree of separation. If each of these 25 people also knows 25 other people, then there's 2 degrees of separation between you and $(25)(25) = 25^2 = 625$ people. If each of these (625) people knows 25 other people, then there's 3 degrees of separation between you and $(25)(25)(25) = 15\,625$ people. You can see how to continue this.

If you play with the numbers a bit, you come up with $43^6 = 6.3$ billion people, which is approximately the population of the earth. In other words, if you know 43 people, and each of these 43 people knows 43 other people, and .... Then there are no more than 6 degrees of separation between you and every other person on the earth.

The first objection everybody has to this calculation usually has something to do with a small tribe on a lost island somewhere that doesn't interact with the rest of the world. Let's agree to ignore small isolated pockets of people. The calculation is still pretty accurate.

The result of this calculation is not unique. That is, if we can assume that each person knows 90 other people, then we have $90^5 = 5.9$ billion people, again approximately the population of the earth. This is no more than 5 degrees of separation. Going the other way, $25^7 = 6.1$ billion. If everybody only knows 25 other people, then there's still no more than 7 degrees of separation. There is no "right" choice, 6 degrees of separation is the popular number.

The biggest problem with all of this is the assertion that someone "knows 43 *other* people." This doesn't often happen. Thirty years ago we could have used chain letters as an example of this assertion. Today we're probably better off using bad jokes on the internet as an example: you receive a bad joke from someone you know (or more likely, after the 50th such occurrence, wish you had never met). This joke is cute enough that you forward it on to a dozen or so people that you know. Within a week, more or less, you receive the same joke from other people that you know. Just where the interconnect network doubled back on itself and got to you again you'll probably never know. The point is that this happens almost every time.

We will approach the simulation of people networks by taking a given number of people and randomly generating a specific number of interconnects. This is an oversimplification in that people's associations tend to group rather than be random. However, it is a useful starting point in that it will let us look at some statistics of how many people know how many people, etc.

Showing a "map" of a 100 000 person network with an average of, say, 20 interconnects per person is not realistic. Instead, we'll illustrate some of the characteristics of these networks on very small networks and just quote the results for the larger networks.

Figure 15.1 shows a 5-person network with 6 interconnects. The five people are shown as small circles with the person referred to by the number in the circle. The positioning of the small circles evenly on a big circle (not shown) and the consecutive numbering of the small circles is completely arbitrary. This is just a convenient way to lay things out. No geometric properties such as proximity of one person to another should be inferred from the positions of the small circles on the page.

In this network, people 1, 2, and 5 have 3 interconnects (friends) each, person 4 has 2 interconnects, and person 3 has only 1 interconnect. This works out to an average of 2.4 interconnects per person. Play with the figure a bit and you'll see that there are a maximum of 10 interconnects, which works out to 3 interconnects per person. Try the same thing with 4 people and you'll get a maximum of 6 interconnects, with 6 people it's 15 interconnects, etc. For very large networks such as the significant fraction of the number of people on the earth, we don't really care about fully interconnected networks such as this because this implies that everybody knows everybody else. However, if you're

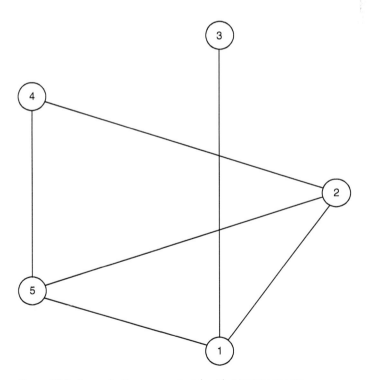

Figure 15.1 Example: 5-person network with 6 interconnects.

curious, the relationship between the number of people and the maximum number of interconnects is

$$\text{max interconnects} = \frac{n(n-1)}{2} \tag{15.1}$$

where n is the number of people.

An important possible property of these networks can be seen by looking at the 9-person, 11-interconnect network shown in Figure 15.2. Although there indeed are 9 people and 11 interconnects, this really isn't a 9-person, 11-interconnect network at all. Persons 1, 2, 4, and 6 form a 4-person, 4-interconnect network and persons 3, 5, 7, 8, and 9 form a 5-person, 7-interconnect network. These are two networks that have nothing to do with each other.

Typical real-world people networks are usually constructed around some guiding rules with a bit of randomness thrown in. The rule could be "members of an extended family" or "fans of Elvis" or "Arizona Amateur Soccer League Players," or co-workers, or neighbors, or,.... A person is usually a member of many more than 1 of these affinity groups. However, backing off to a far enough

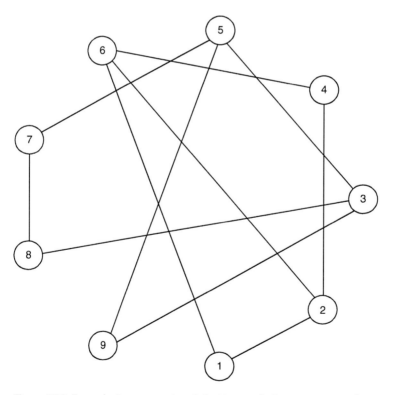

Figure 15.2 Example: 9-person network that is actually 2 separate networks.

distance and looking at the networks for such groups, tends to give us a picture that looks pretty random. If we're dealing with hundreds of million or even billions of people, there certainly will be isolated subnetworks that exist for no good reason whatsoever. Remember that the networks that are ultimately of interest to us are untenable to draw completely, such as the "6 degrees of freedom" network which consists of 6 billion people and many, many interconnects.

Propagation Along the Networks

Following something along the interconnect paths in our networks can have many physical interpretations and as many sets of rules. We could, for example, be interested in what happens with chain-letter money deals. In this situation you mail $1 to either everybody you know or some fixed number of people (assuming that everybody has at least this fixed number of acquaintances). In this case you'd be interested in how much money ultimately gets back to you.

Another interest would be to follow the propagation of a contagious disease. Here we need some contagion rules. A simple starting point is to assume that you see everybody you know today, and tomorrow they all have the disease (and you're dead or recovered and no longer contagious). This is pretty unrealistic, but it's an interesting extreme case with which to start studying propagation properties such as "how much time do we have to develop a vaccine or cure," and "does everybody die?" Anticipating, the answer to the latter question is *no* because of the isolated sub-networks described above.

When generating a random network for some given number of people, there are two ways to approach the randomness. First, we could fix the number of interconnects per person and then scatter the "far end" of each interconnect randomly. Or, we could fix the total number of interconnects and then scatter both ends randomly, only requiring that there be no duplications.

We chose the latter approach because it leads to more realistic results. The number of interconnects per person, by just about any set of affinity rules, is certainly not fixed. However, by fixing the total number of interconnects, we can "play" with this total number and study the implications to various consequences.

Our first example is a 10 000-person network with 50 000 interconnects. Although 50 000 may seem like a large number, remember that for a 10 000-person network the maximum number of interconnects is

$$\frac{n(n-1)}{2} \approx \frac{n^2}{2} \left(\text{for } n \text{ large} \right) = \frac{10000^2}{2} = 50\,000\,000 \tag{15.2}$$

50 000 interconnects is only .1% of the possible number of interconnects for this network.

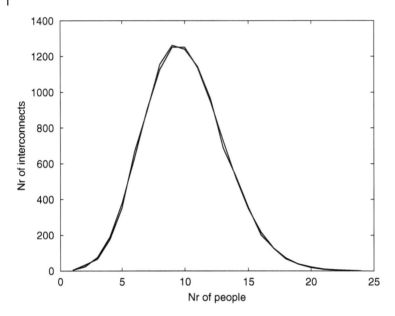

Figure 15.3 10 000 Person, .1% of maximum random interconnects, distribution, superimposed on a Poisson distribution.

A network satisfying these numbers was generated using Monte Carlo simulation techniques. The average number of interconnects per person = 10.0 (for 10 000 people, 50 000 interconnects at 2 people per interconnect), the average number of interconnects per person must be 2 (50 000/10000) = 10.0. Had the simulation calculated a different number, it would have had an error somewhere.

The standard deviation of the above average = 3.14. Note that $3.14 = \sqrt{10}$, and $\sigma = \sqrt{\mu}$ is a characteristic of a Poisson distribution.

Figure 15.3 shows the distribution of interconnects per person, superimposed on a Poisson distribution with a mean of 10.0. As you can see, the fit is excellent; the distribution of interconnects per person is indeed a Poisson distribution.[1]

The axes of Figure 15.3 may not be clearly explained by their labels. The simulation was run, as stated above, with 10 000 people, but the horizontal axis, labeled "Nr of people" only runs from 1 to 25. This axis is not listing all the people, it is listing how many people the interconnects have specified by the vertical axis. For example, 10 people have approximately 1200 interconnects each, 15 people have approximately 200 interconnects each, etc. This is a

---

1 Again, remember that these are both discrete, not continuous, distributions. The graph is drawn with continuous lines for ease of viewing.

discrete distribution and if we add up the Nr of Interconnects numbers for each Nr of People, the result of course would be 10 000.

It's not hard to imagine a town with a population of 10 000, each person knowing (on the average) 10 people. We could, in principle, draw the same type of grid as shown in Figure 15.1. Doing this is not practical, but we should note that our new, much larger grid, conveys the same information as the simple grid. Using this generated grid, we can examine how some information or a disease propagates in this town.

If we assume that each person "touches" every person on their interconnect list each day and that this touch transfers a contagious disease each day, then the disease spreading problem is identical to the "degrees of freedom" problem; the statistics of the number of days it takes to infect everybody is identically the statistics of the "degrees of separation" problem. Also, it is very similar to information spreading on a computer social network. In this latter case, "friends" are the interconnects.

Using the same network as above (10 000 people, 50 000 interconnects) and the absolute disease spreading rules described above, we can simulate the spread of disease through our little town. The results are shown in Table 15.1. The first person infected spread the disease to 9 other people. Each of these people spread the disease to other people, etc. In eight days, everybody gets infected and then dies.

Repeating this simulation using a network of 10 000 people and only 10 000 interconnects, we get a different result (Table 15.2). Even though the infection is guaranteed to propagate through the network with 100% probability, the number of people infected never rises past 7972.

Table 15.1 Hypothetical infection spread in highly connected network, 100% infection probability.

Day	Healthy	Infected	Dead
0	10 000	0	0
1	9999	1	0
2	9990	9	1
3	9892	98	10
4	8934	958	108
5	3427	5507	1066
6	15	3412	6573
7	0	15	9985
8	0	0	10 000

Table 15.2 Hypothetical infection spread in lightly connected network, 100% infection probability.

Day	Healthy	Infected	Dead
0	10 000	0	0
1	9999	1	0
2	9998	1	1
3	9996	2	2
4	9990	6	4
5	9975	15	10
6	9948	27	25
7	9897	51	52
8	9788	109	103
9	9583	205	212
10	9170	405	417
11	8452	727	822
12	7333	1118	1549
13	5889	1444	2667
14	4372	1517	4111
15	3195	1177	5628
16	2549	646	6805
17	2246	303	7451
18	2119	127	7754
19	2059	60	7881
20	2028	31	7941
21	2020	8	7972
22	2013	7	7980
23	2010	3	7987
22	2009	1	7990
23	2009	0	7991

What is happening here is that in a town where each person knows, on the average, only two other people, it's pretty likely both that there are islands of interconnects that are or become isolated from the each other or people contracting the disease and then becoming noncontagious and breaking interconnects. This is, on a larger scale, the situation depicted in Figure 15.2. Repeating this same simulation over and over again gives slightly different results each time; but this is telling us about the nature of random-association networks as much as about disease propagation.

Now, what if it's not guaranteed that an infected person infects every person they know? This is very realistic. For one thing, you don't usually see everybody you know every day. Second, if the disease is spread by coughing or sneezing, you simply might not pass it to everybody you see. Some people will have higher levels of immunity than others and are not as easily infected. And finally, if transmitting the disease requires intimate contact, this usually translates to the disease only being passed to a small percentage of the people that, in some sense, a person knows.

Figure 15.4 shows the results of the same simulation couched in terms of the total number of people dying as a function of the probability of passing the infection to someone you know (original number of interconnects). Remember that this is a very simple-minded system. You get the disease on day one; you pass it to everybody you know (with a given probability of passing it) on day two, on day three you're either dead or cured and are no longer contagious. Whether you are dead or cured is certainly very important to you, but is of no consequence to the propagation of the disease: no longer contagious is no longer contagious.

When the probability of infecting a contact is low enough, the sick person dies or recovers before they get another chance to infect that contact. When all the connections to a contact or connected group containing that contact get

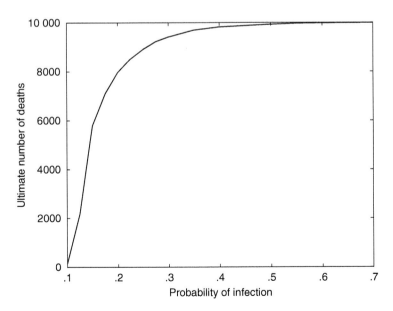

Figure 15.4 Number of deaths versus probability of infection.

broken (by death or recovery) the contact or group is "safe." For a low enough infection probability a significant portion of the population is safe and the infection dissipates.

If we were to repeat the calculation for a more densely populated network, the curve would have the same general shape as that of Figure 15.4, but the numbers would be shifted a bit. In particular, we need a lower probability of infection for the more densely populated network in order to see the "cutoff" of disease spread.

We could add more realism to the network and contagion rules to get curves that look more like real-life situations. A city's network is not just a random connection grid; there are "affinity groups" with many interconnects, these groups are sparsely tied to each other, etc.

Some Other Networks

Diseases aren't the only thing that propagate along networks. In today's world, information transfer along networks is a major factor in all facets of communications.

The telephone system is a network of telephones and interconnects. The telephone system is of course a "managed" network in that when you want to talk to your friend the interconnection is set up; your telephone line (interconnect) goes to a central office where the electrical version of your conversation is converted to digital information. This digital information then gets to "time-share" on a large central interconnect to some central office near the person on the other end of your call. There are possibly several routing points that have been passed through.

The cellular telephone systems, from a network point of view, are made up of base-stations that are "central offices" that connect the wired, or land-line, telephone system to radio connections to your cellular handset. The cellular system is very intelligent in that as you move in and out of range of different base stations, your call is "handed over" to the closest base station without you ever knowing that this hand-over has occurred.

The internet is an interconnection of digital networks. On the internet, your information is broken into short packets, which then travel, along with other folks' messages, on the "information superhighway."

In all of these information handling networks, there is always the possibility that a piece of a message will get corrupted. A lightning flash near some buried cable can, for instance, create a pulse of signal along the cable that will change some of the digital pulses. There are some very sophisticated systems for error detection and correction along these networks. Understanding these systems is an education in itself.

Neighborhood Chains

The disease spread example above was based on a network where any two "friends" are equally likely. In a real town, it's more likely that you know someone who lives near you than someone who lives all the way on the other side of town. To look at this, we must assign an "address" to every person. We will assign, on a grid, a unique location to each person.

You could substitute, for example, "is related to you" for "lives near you" and a family tree for address distances without overall patterns changing very much. In other words, using physical distance is just one example of many types of affinity relations. When you mix physical distance with affinity relations that spread all over town with membership in the community choir, etc., you start forming networks that simply look like the random ones discussed earlier in this chapter.

In the real world, we have neighborhoods building up towns, a bunch of towns with low population density between them building up a region, etc. As a simple example just to show some patterns, we'll look at a region with 225 people. This is a very small number of people for simulating real situations but it's as big a population as we can show graphically on a single sheet of paper with readable resolution.

Figure 15.5 shows our town. Each "o" represents the home address of a person. The homes are on a simple square grid. This is not realistic, but it saves us from blurring the picture with patterns of homes. Figure 15.6 shows this same town with 200 randomly placed interconnects. This works out to each person knowing, on the average, 400/225 ~ 2 other people. Again, this is an unrealistically small number but it allows for a viewable sketch.

Near the upper left corner of this figure you'll see a person who knows only one other person. There are also missing circles, i.e. people who don't know anybody. These occurrences are very unlikely when there is a more realistic (i.e. higher) number of interconnects per person, but not at all unlikely in this situation.

Figure 15.7 shows what can happen when the random selection of interconnects is constrained. To generate this sketch, the random selection of interconnects has been modified to include a scaling that forces a high probability of creating an interconnect if the separation between the two homes is small and a low probability if it's large. There are 200 interconnects, just as in the previous figure. However, these two figures look very different. To the extent that this approximates some real situations, you can see that it's not that easy for contagious diseases to get to everybody. If the person on either end of the interconnect between otherwise isolated neighborhoods is away that week, or recovers (or dies) from the disease before he gets to visit his friend at the other end of the interconnect, the chain is broken.

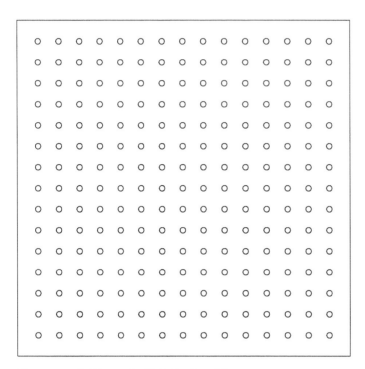

Figure 15.5 Grid layout for "Neighborhoods."

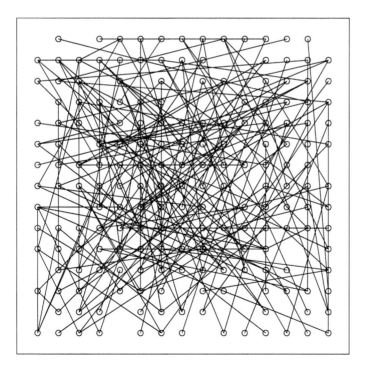

Figure 15.6 Random interconnects in grid layout.

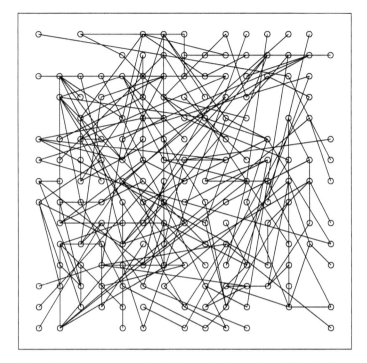

Figure 15.7 Interconnects in grid layout having preferred "Nearby" interconnects.

Chain Letters

An interesting network in which information travels both forward and backward is the chain letter. There are many versions of the chain letter, so this description might not agree exactly with what you've seen before.

Your interaction with a Chain Letter begins when you get a letter in the (e) mail. The letter has a list of six names and addresses. Your instructions are to:

1) Mail one dollar to each of the names and addresses on the list
2) Remove the bottom name and address from the list and add your name and address to the top of the list
3) Mail 6 copies of the letter to people you know, excluding any of the 6 names on your incoming list.

There are 3 ways to interpret how this will work. First, we can go with the original 6 degrees of freedom assertion that, in this situation, means that each person mails letters to 6 people who have never yet gotten this letter. Second, we can make the more reasonable assumption (as we did in the disease propagation discussion) that the first scenario never happens and you start getting repeat copies of the letter (but you throw them away). Third, each time you receive the letter you repeat the process, mailing 6 copies to 6 people that you haven't mailed the letter to before. In this latter case, eventually you'll come to a stop because you don't know an infinite number of people.

In all of these cases, we have to be able to weigh in the factor that some percentage of the letter recipients will simply throw the letter away. This is analogous to the infection probability of the disease network.

Figure 15.8 shows the ideal chain letter as described above. You (person X) receive a letter with names A, B, C, D, E, and F. You cross out name F from the

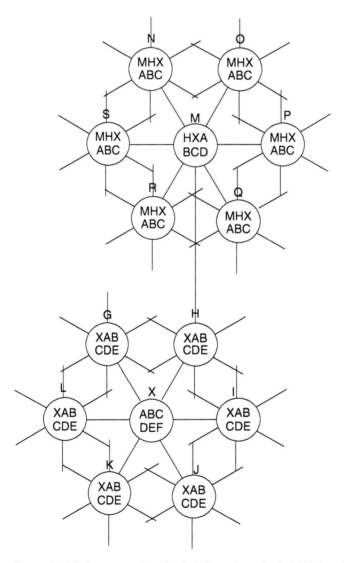

Figure 15.8 Early propagation of a chain letter, hypothetical 100% participation.

bottom of the list and add your name to the top of the list. You mail the letter to G, H, I, J, K, and L.

We'll just follow the path to person H. He crosses out E, adds H, and mails the letter to 6 people, one of whom is person M. Person M sends out the letter containing the names M, H, X, A, B, C to N, O, P, Q, R, and S, and so on.

After six such mailings, your name is crossed off the bottom of the list and the future of this chain is of no further interest to you.

At level 1, each of the 6 people sends $1 to you, giving you $6. At level 2, each of the $6^2 = 36$ people sends you $1, giving you $36 + $6 = $42. If everything goes perfectly, you ultimately receive

$$\$6 + \$36 + \$216 + \$1296 + \$7776 + \$46\,656 = \$55\,986$$

This is not bad for a $6 investment (not counting postage). You have multiplied your money by a factor of $55\,986/6 = 9331$.

There are several things to consider here before you participate in one of these chains. We'll only consider the probabilities involved. There might also be legal issues; this could be considered a scam.

As was mentioned above, there have to be 55 986 people involved, only 6 of whom know you. Every one of these people has to participate fully (look out for the ones who send the letters forward but never send the money back).

Sadly enough, joining a chain letter is not a good way to invest your money.

At 6 levels, we needed 55 986 people participating (actually, 55 987 if we count you). What if this has been going on for a while and you receive the letter at level 3, that is 3 levels further along than is shown in Figure 15.8? For the chain to complete, from your perspective, over 12 million people must participate. Three more levels and we're up past 2.6 billion people. That's about 1/3 the population of the earth! And remember that we're still maintaining the restriction that each one of these people know only the 6 people on the list – the 6 people that they send their letters to.

The conclusion thus far is that you should never jump onto even an idealized one of these chains once a few levels have passed by. Now let's get more realistic: keep the restriction on the people not knowing each other but look at what happens when some of the people throw their letter away. Let's assume that you started the chain, so we have identically the situation depicted in Figure 15.8.

Consider a person who just joined the chain at level 6. If he never mails you 1$, you're just out a dollar. On the other hand, person A is one of your original 6 people. If she trashes your letter, then fully 1/6 the chain is broken and you're out that fraction of your winnings, that is $9331.

While getting the full $55 986 from the chain letter is unlikely, getting some money isn't difficult. For a 50% participation, the expected amount received is about $1100, which isn't bad for an hour's work addressing envelopes. However, the curve falls quickly and at 20% participation the amount falls to about $12.

This is beginning to look like a waste of time. On top of all of this, we are still dealing with the totally ridiculous assertion that we could find over 55 000 people, all of whom would mail letters only to people who would only receive this one letter – especially when each person only sees the 6 names on the list they receive, so they couldn't even try to avoid duplication.

The more realistic chain, that is the chain where we acknowledge that people tend to know each other in groups rather than in random links, will behave pretty much like the chain letter above but with a random probability of someone trashing the letter. Also, there's a random chance that the chain will work its way back to you and you trash the letter, but this isn't computationally different from the chain reaching a new person who trashes the letter.

Comments

Chain investing schemes are an example of how the mathematics of networks can be used to entice people to give up their money. On the other hand, there is real value in studying these networks, not just to understand disease propagation (which alone would be reason enough). Working with telephone, or data, or power network simulations is also very useful because it lets designers study network capacity and failure probabilities. Since resources are always limited, it's important to minimize the probability that a local event such as an earthquake or a fire can shut down an entire network because the remaining branches either don't make the necessary connections or are themselves overloaded and fail. This has happened to several pieces of the US power grid and it's not pretty. The internet – on the other hand – has significant rerouting capabilities. Often, totally unawares to us, different pieces of the web page heading our way are actually traveling along very different paths.

16

Introduction to Frequentist Statistical Inference

Introduction

In previous chapters we introduced the distinction between frequentist and Bayesian concepts of probability, and in Chapters 11 and 12 developed the basics of Bayesian statistics. There are many situations where the frequentist approach is appropriate, and for that matter most discussions of "Statistics" or "Statistical Inference" refer to frequentist statistics.

Both frequentist and Bayesian statistics are enormous fields of knowledge. In this introductory book, we have only scratched the surface of Bayesian statistics; similarly, here we will only scratch the surface of frequentist statistics.

Sampling

Suppose we have a population of something that we want to learn about. A good example is the distribution of lifetimes of light bulbs that some manufacturer produces. This example is mathematically equivalent to the average weight of bottles of juice, the average amount of sugar in bottles of soda, etc. A related but not quite identical example is the incidence rate of broken cashew nuts in jars (of cashew nuts), or of nails without heads in boxes of nails. All of these examples are important to manufacturers (and purchasers) of these items.

Related to the above is the problem of measurement. No measurement instrument is perfect. In addition, measurements of weights and volumes tend to vary with ambient temperature and humidity, neither of which are easy to control precisely. A large manufacturer of some product will probably have multiple manufacturing lines running, each with their own measurement machines. Sometimes these machines are right next to each other in a factory, and sometimes they're distributed in factories around the world. There is an

Probably Not: Future Prediction Using Probability and Statistical Inference,
Second Edition. Lawrence N. Dworsky.
© 2019 John Wiley & Sons, Inc. Published 2019 by John Wiley & Sons, Inc.
Companion website: www.wiley.com/go/probablynot2e

issue of correlation of the different measurement systems. Putting all of this together, if we were able to take one bottle of juice and measure its weight on every one of the manufacturer's scales in every factory, several times a day, in every season of the year, we'd find a distribution of results.

In practice, both of the above phenomena occur simultaneously. That is, every bottle of juice is slightly different from the others, and no two measurements of any single bottle of juice repeat exactly. How do we know what we have?

The first problem to address is the sampling problem. We don't want to run every light bulb until it burns out in order to see how long it lasts. We don't want to inspect every nail to see if it has a head. We would like to look at a small representative group of items and infer the information about the total population.

The last sentence above contained two words that each contain a universe of issues. What do we mean by *small*, and what do we mean by *representative*? We will be able to define *small* mathematically in terms of the size of the entire population, the size of confidence intervals, etc. *Representative*, on the other hand, more often than not depends on judgment and is not easy to define. It is the factor that leads many surveys about peoples' opinions and preferences astray.

Let's start with an example that doesn't involve people. You're a bottled juice manufacturer and you want to know the probability distribution function (PDF) of the weights of the bottles of juice without weighing every bottle. You could start out by taking a standard weight and using it to calibrate every scale in all of your factories, but the time involved in shipping this weight to these factories around the world is large as compared to the time it takes for scales to drift out of calibration. Instead, you'll try to develop a set of standard weights and have a few of each located in every factory. You could have a revolving system, with some fraction of the weights from each factory being brought together periodically and checked against one master, or "gold standard" weight.

How often do you need to calibrate your scales? This depends on the scales, the temperature and humidity swings in factories, etc.

Once you've calibrated the scales, you have to set up a sampling procedure. You could take, say, 10 bottles of juice at the start of each day and weigh them. The problem here is that the juice bottle filling machines heat up over the course of a day and the average weight at the end of the day isn't the same as the average weight at the beginning of the day. Modify the procedure and weigh 3 bottles of juice from each filling machine three times a day. This is getting better, but the sampling process is growing. As you can see, this is not a trivial problem.

The problems involved with surveying people are much more involved. Suppose we want to know the average number of hours that a person in New York City sleeps each night and what the standard deviation of this number is.

We could start with a mailing list or a telephone list that we bought from, say, some political party. It's probably reasonable to assume that democrats, republicans, libertarians, anarchists, and communists all have statistically similar sleep habits. Or do they? We don't really know, so we're already on thin ice buying this mailing list.

We know that we won't get responses from everybody we call, If we telephone people, then we could pick up a bias; people who aren't sleeping well are probably crankier than people who are sleeping well and are more likely to hang up on us. Also, should we call mid-day or early evening? People who work night shifts probably have different sleep patterns than people who work day shifts, and we'll get disproportionate representation of these groups depending on when we call.

Lastly, how well do most people really know how many hours they sleep every night? Our memories tend to hold on to out-of-the-ordinary events more than ordinary events. If you had a cold last week and didn't sleep well, you'll probably remember this more than the fact that you slept perfectly well every night of the previous month. Also, sleep studies have shown that many people who claim that they "didn't sleep a wink last night" actually get several hours of sleep in short bursts.

Finally, although it's hard to imagine why people would want to lie about how well they're sleeping, in many situations we simply have to consider the fact that people are embarrassed, or hiding something, and do not report the truth. Sexual activity and preference studies, for example, are fraught with this issue. Also, mail or email survey responses are usually overrepresented by people with an axe to grind about an issue and underrepresented by people who don't really care about the issue.

The bottom line here is that accurate and adequate sampling is a difficult subject. If you're about to embark on a statistical study of any sort, don't underestimate the difficulties involved in getting it right. Read some books on the subject and/or consult professionals in the field.

One last comment about sampling: many people and companies get around the problems of good sampling by simply ignoring the problem. An example of this is often seen in the health food/diet supplement industry. Valid studies of drugs, including health foods and "disease preventative" vitamins and diet supplements take years, and include many thousands of people. There are so many variables and issues that there really isn't a simple short cut that's valid. Instead, you'll see bottles on the shelves with testimonials. Good luck to you if you believe the claims on these bottles. For some reason, many people will believe anything on a bottle that claims to be Chinese or Herbal, but suspect the 10-year, 10 000 people, FDA study that shows that the stuff is, at best, worthless. In the case of magnetic or copper bracelets or cures for the common cold, this is relatively harmless. In the case of cancer cures you could be signing your own death certificate.

Returning to the mathematics of sampling, the most bias-free sampling technique is random sampling. In the case of the bottled juice manufacturer, use a computer program that randomly picks a time and a juice bottle filling machine and requests a measurement.[1] Some insight must be added to prescribe the average number of samples a day, but this can usually be accomplished with a little experience and "staring at" the data.

An assumption that we must make before proceeding is that the statistics of the population of whatever it is we're studying is not changing over the course of our study. Using the juice bottle example from above, suppose we're weighing 25 bottles of juice at randomly chosen times each day, but a worker (or a machine glitch) resets the controls on the filling machine every morning. All we're accomplishing in this situation is building a history of juice bottle weighing. The information from Monday is of no value in telling us what we'll be producing on Tuesday.

Sample Distributions and Standard Deviations

Consider a factory that produces millions of pieces of something that can easily be classified as good or defective. The example of nails, where nails with missing heads are defective parts, fits this category. There is a probability p of the part being defective and $(1-p)$ of the part being satisfactory to ship to a customer. The manufacturer plans to allow the shipment of a few defective parts in a box along with a majority of good parts. The customers will do the final sorting: they'll pick a nail out of the box and if it's bad, they'll toss it away. So long as the fraction of defective parts doesn't get too large, everybody is happy with this system because the quality control/inspection costs are very low. In principle, the manufacturer could inspect every part and reject the bad ones, but this would be very expensive. On the other hand, if p is too large the customers will get annoyed at the number of bad parts they're encountering and will switch to another manufacturer the next time they buy their nails. It is important for the manufacturer to know p accurately without having to inspect every part.

The manufacturer decides that every day he will randomly select n parts from the output of his factory and inspect them. If the parts are fungible and are sold in boxes, such as nails, then a box of n nails is a likely sample. On Monday he finds x bad parts and calculates the probability of a part being bad as

$$\bar{p} = \frac{x}{n} \tag{16.1}$$

---

1 Even this can go wrong in the real world. For example, a well-meaning worker might try to make his day simpler by grabbing a dozen bottles of juice in the morning, putting them near the scale, and then weighing one of them each time the computer buzzes.

We want to know p, the probability of a bad part (or equivalently, the fraction of bad parts) that the factory is producing. We have not measured p. We have measured the fraction of bad parts in a random sample of n parts out of the population distribution. We called this latter number $\bar{p}$ (usually read "p bar") to distinguish these two numbers. p is the fraction of bad parts in the population, $\bar{p}$ is the fraction of bad parts in the sample of the population (usually simply referred to as the sample).

The manufacturer would like these two numbers to be the same, or at least very close, so that she can infer p from $\bar{p}$. We can get a feeling for the problem here by looking at using the binomial probability formula. Suppose that $p = .15$, so that $(1 - p) = .85$. This defective part fraction is unrealistically large for a commodity product such as a box of nails, but is good for an example.

We would expect to find about 15% bad parts in a sample. The probability of finding exactly 15 bad parts in an $n = 100$ sample is, using the binomial probability formula,

$$C\binom{n}{k} p^k (1-p)^{n-k} = \frac{100!}{85!15!}(.15)^{15}(.85)^{85} = .111 \tag{16.2}$$

There is about an 11% probability that $\bar{p}$ will correctly tell us what p is.

Figure 16.1 shows the probability of getting various fractions of bad parts in a random sample of 100 parts. As can be seen, the mean of the curve is at .15. The curve looks normal. This shouldn't be a surprise because several chapters ago we showed how a binomial probability will approach a normal curve when the n gets big and p or q isn't very large. The standard deviation is .036. The 95% confidence interval is therefore .072 on either side of the mean. This means that we can be 95% confident that we know the fraction of bad parts being produced to within .072 on either side of the nominal value (.15).

Also shown in Figure 16.1 is the results of doing the same calculation for $n = 200$. Again, the mean is .15, but now the standard deviation is .025. Since the standard deviation has gotten smaller, we have a smaller confidence interval and hence a better estimate of the fraction of bad parts the factory is producing. If we take the ratio of the two standard deviations, we find that the larger number is 1.42 times the smaller number. This is, not coincidentally, a ratio of $\sqrt{2}$. The standard deviation of mean of the sample distribution falls with the square root of the size of the sample.

The standard deviation of $\bar{p}$ is given by

$$\sigma = \sqrt{\frac{p(1-p)}{n}} \tag{16.3}$$

An interesting point about the above equation: p is the probability of a defective part, so $q = 1 - p$ is the probability of a good part. The equation has the

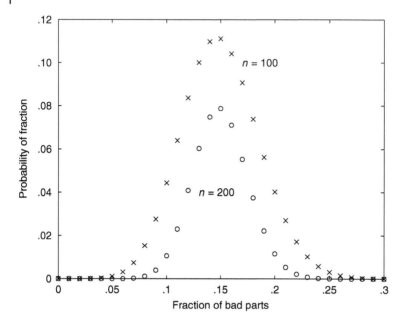

Figure 16.1 Bad parts sampling distributions for $n = 100, 200$.

product of these two probabilities, i.e. its value does not change if we put it the value of q rather than the value of p:

$$p(1-p) = q(1-q) \qquad (16.4)$$

The standard deviation of the fraction of good parts in a sample of size n is the same as the standard deviation of the fraction of defective parts in a sample of the same size.

Estimating Population Average from a Sample

Next let's look at a population with a normal distribution of some property. The weights of bottles of juice is a good example. A factory produces bottles of juice with an average weight (mean of the distribution) w and a standard deviation σ.

Again, assume that the measurement system (scales) are ideal.

We randomly select n bottles of juice and weigh them. Naturally, we get n different results. Figure 16.2 shows a histogram of a possible distribution of 100 measurements of a population with $W = 10.00$ and $\sigma = .25$. The average of these numbers is 10.003 and the standard deviation is .256. This sample seems to emulate the true population in average and standard deviation,

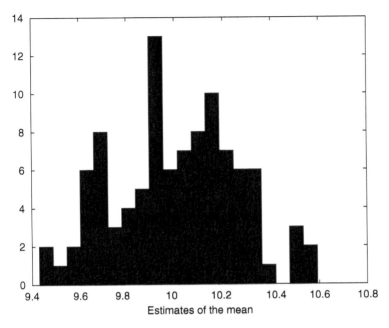

Figure 16.2 Juice bottle sampling data, 100 measurements.

but the histogram of the data certainly does not look like a well-behaved normal distribution.

In order to see how well averaging the results of a 100 measurement sample predicts the mean of the population, we'll repeat the procedure 1000 times and graph the results (Figure 16.3). The average obtained is 9.9998 and the standard deviation is .0249. Remember that this average is the mean of a very large number of 100 measurement averages. This mean is a random variable itself, with an expected value of W and a standard deviation of

$$\frac{\sigma}{\sqrt{n}} \tag{16.5}$$

The average of n samples is the best estimate of the mean of the population, and the standard deviation of this average decreases as n gets larger. In other words, to get a very good idea of W, take as many randomly sampled measurements as you can and average them. Unfortunately, since the standard deviation of your result is only falling with the $\sqrt{n}$ rather than with n itself, improvement gets more and more difficult as you proceed. Going from $n = 25$ to $n = 100$, for example, will cut the standard deviation in half. However, to cut it in half again, you would have to go from $n = 100$ to $n = 400$.

If the measurement system has some variation, then we must somehow take this variation into account. Suppose we took one bottle of juice and measured

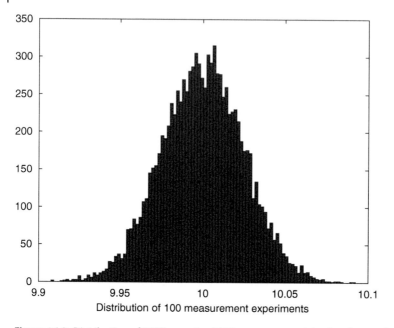

Figure 16.3 Distribution of 1000 repeats of 100 measurement juice bottle sampling.

it many times by randomly selecting the different scales in our factory(ies), different times of day, etc. The average of all of our results is our best estimate of W, the weight of this bottle of juice. If, instead of a bottle of juice, we used a standard weight, then we would calibrate our system so that this average of all of our results is indeed the weight of the standard.

For a measurement system standard deviation σ_m and a population standard deviation σ, the total variation we have to deal with is

$$\sigma_{tot} = \sqrt{\sigma^2 + \sigma_m^2}$$

assuming that these two standard deviations are independent (of each other).

The formulas above assume that the population you're studying is normal, or nearly normal. This is usually a safe assumption, but be a little suspicious.

Also, these formulas assume that you know σ, the standard deviation of the population and, if necessary σ_m, the standard deviation of the measurement repeatability. What if you don't know these numbers? The easiest way to approximate them is to take the standard deviation of the n samples. Just in case some confusion is building here, remember that the standard deviation of the n samples in a measurement set must approach the standard deviation of the population as n gets large and eventually approaches the size of the population itself. The standard deviation that gets smaller with increasing n is not something you usually measure: it's the standard deviation of the averages of a large number of n-sized measurement sets.

The Student-T Distribution

It's easy to picture a set of a few hundred measurements from a population having a standard deviation that approximates the standard deviation of the entire population. But what about a dozen measurements, or five, or even just two? With one measurement, we can't calculate a standard deviation; two measurements is the smallest set of measurements that we can consider when talking about approximating a standard deviation. Remember that the calculation of standard deviation has the term $n - 1$ in the denominator of the formula, and this makes no sense when $n = 1$. The $n = 2$ case is the smallest size measurement set possible, and is referred to as a "one degree of freedom" case.

The best way to look at what happens with small values of n (small samples) is to *normalize* the distribution curves, that is, to first shift the curves on the horizontal axis until they all have the same mean, then to divide by $\frac{\sigma}{\sqrt{n}}$. In texts and handbooks it is common to define a "standard" random normal variable z,

$$z = \frac{\bar{x} - \mu}{\dfrac{\sigma}{\sqrt{n}}} \tag{16.6}$$

This transformation makes it possible to examine the properties of all normal distributions by just studying one normal distribution.

We'll keep the letter σ as the population standard deviation and call the standard deviation calculated from an - sized sample s. Then we'll define a new normalized variable as t,

$$t = \frac{\bar{x} - \mu}{\dfrac{s}{\sqrt{n}}} \tag{16.7}$$

Figure 16.4 shows the t distribution for $n = 2$, $n = 4$, $n = 7$, $n = 31$ along with the variable z. The first four curves correspond to 1, 3, 6, and 30 degrees of freedom, respectively.

The T curves are called "Student-T" distributions, named after the mathematician who developed them and used the pen-name Student. As n increases, the Student T curves look more and more normal until at about $n = 30$ (and larger) they're virtually indistinguishable. However, for low values of n (small samples), the curves are flatter and wider than normal curves. Table 16.1 shows the x value for the 97.5% point at various values of n. Since this is a symmetric distribution the 95% confidence intervals are twice these x values.

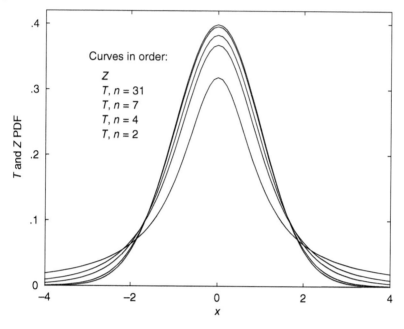

Figure 16.4 Student-T distributions, $N = 2, 4, 7, 31$.

Table 16.1 Student-T distribution, 97.5% confidence interval limits (about mean) versus n.

Deg. of Freedom	n	x
1	2	12.7
2	3	4.30
3	4	3.29
4	5	2.77
5	6	2.58
7	8	2.37
11	12	2.20
15	16	2.14
16	2	2.10
23	24	2.06
29	30	2.04
Z	—	1.96

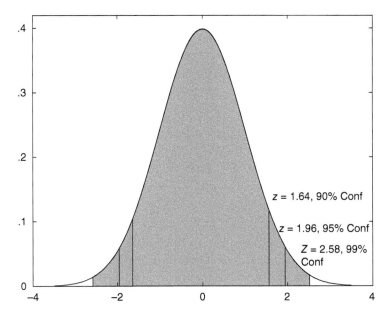

Figure 16.5 Standardized normal distribution showing confidence interval points polling statistics.

For $n = 2$, the confidence interval is 25.4. Compare this to the high n values, or the z value, where the confidence interval is approximately 2. The Student T distribution is telling us that when we have very little data and no prior information, we do not know very much!

If distributions had personalities, we would say that the Student-T distribution is a very suspicious PDF. When you don't have too much information (i.e. a small sample) then the calculation of s is only a poor approximation of σ, and the confidence interval is *very* large. As n increases, the confidence interval decreases and it approaches the value predicted by the normal (z) curve.

What if the original population (e.g. the juice bottle weights) is not normally distributed? It turns out that it doesn't matter: the Central Limit Theorem guarantees that the distribution created by taking the average of n experiments many times and calling this average our random variable will be (essentially) normally distributed. The conclusions reached above are therefore all valid.

Converting calculations to the z or t variables offers us the convenience of seeing everything in a common set of parameters. That is, we are transforming all normal distribution problems to that of a normal distribution with a mean of 0 and a σ of 1. We know that the range between $z = -1.96$ and $z = +1.96$ is our 95% confidence interval, etc.[2] Figure 16.5 shows the 90%, 95%, and 99% confidence interval points.

---

2 It is common practice to refer to the normal distribution's 95% confidence interval as $\pm 2\sigma$ even though $\pm 1.96\sigma$ is more accurate.

If you are reading this book any time when there's a local or national election on the horizon, you are no doubt satiated with pre-election polling results in the newspapers and the TV and radio news reports. In a city of 1 million voters, for example, a pollster samples 1000 people (hopefully randomly selected) and finds that 52% favors candidate A. This doesn't seem like a large margin. Exactly how comfortable should candidate A feel the night before the election? This falls back on seeing how confident we are that the sample probability of 52% of the voters voting for candidate A represents the entire population of 1 000 000 voters.

We estimate the standard deviation of the sample data as

$$s = \sqrt{\frac{\bar{p}(1-\bar{p})}{n}} = \sqrt{\frac{.52(1-.52)}{1000}} = .0158 \qquad (16.8)$$

For a 95% confidence factor, we need 1.96 standard deviations, or (1.96) (.0158) = .031. Our confidence interval on the voting is therefore $.52 \pm .031$. We are 95% confident that candidate A will get between about 49% and 55% of the votes. Remember that if we had taken more samples, n would be larger and s would be smaller than the values shown above; the confidence interval about the 52% result of the survey would have been smaller, and we could be more comfortable with the prediction.

The above example, while mathematically correct, can easily be misinterpreted. The 95% confidence window about the 52% sample result is $\pm.031$. However, candidate A doesn't really care about the + side of things, his only concern is that the actual vote count might give him less than 50%. We should therefore ask just what confidence factor we have of our candidate winning. We need a confidence interval on the low side of $.52 - .02$ to keep him in winning territory.

We need to calculate

$$.02 = .0158z \qquad (16.9)$$

Or,

$$z = \frac{.02}{.158} = 1.27 \qquad (16.10)$$

From a spreadsheet or a table, we see that this is just about an 80% 2-sided confidence factor, i.e. there is a 10% probability of getting less than 50% of the vote and a 10% probability of getting more than $.50 + .02 = 52\%$ of the vote. Since the relevant information for our candidate is only the low side, he has a 90% confidence factor of winning. This isn't a guaranteed win, but it's not too bad for the night before the election.

Did a Sample Come from a Given Population?

Another use for confidence interval calculations is to determine how likely it is that a sample data set really came from a random sampling of some population.

As an example, consider left-handedness. The national incidence of left-handedness is about 12% of the total population. In a randomly selected group of 33 people, we should expect to find about 4 left-handed people. But what if we stand on a street corner, query the first 33 people we meet, and find 7 left-handed people? Are we within the expected bounds of statistical variation or should we be suspicious about this sample?

We start by calculating the sample standard deviation of the probability,

$$\sigma = \sqrt{\frac{p(1-p)}{n}} = \sqrt{\frac{.12(1-.12)}{33}} = .057 \tag{16.11}$$

We would expect $\bar{p}$ to be within $1.96s$ of p (on either side) for a 95% confidence factor. Calculating this number,

$$p + 1.96s = .12 + .112 = .232 \tag{16.12}$$

From the sample data, $\bar{p} = \frac{7}{33} = .212$. Therefore, our sample data is within the confidence interval. To be within this confidence interval, we can therefore conclude that finding 7 left-handed people in a group of 33 isn't suspicious, but there isn't a big margin to this conclusion.

There's much, much more to the field of frequentist statistical analysis than has been presented. However, as we said at the beginning of this chapter, this is primarily a book on probability, so we'll stop here. Hopefully we've conveyed the flavor.

A Little Reconciliation

Neither frequentist nor Bayesian decision making approaches ("did this XXXX come from that population?") has rigid decision making calculations. This is because the question being asked, "Is 7 left-handed people in a group of 33 believably a random occurrence or not?" is not a well-defined question. What does "believably" or the term used previously, "suspicious," actually mean? The best we can do is to have guidelines, i.e. rules-of-thumb. In terms of these rules of thumb, it is interesting to compare/contrast the two approaches.

Figure 16.6 shows the results of a Monte Carlo simulation. The simulation looked at 200 000 iterations of groups of 33 people with each person having a 12% probability of being left-handed. This Monte Carlo simulation, while inefficient numerically, has the advantage that there are no assumptions made

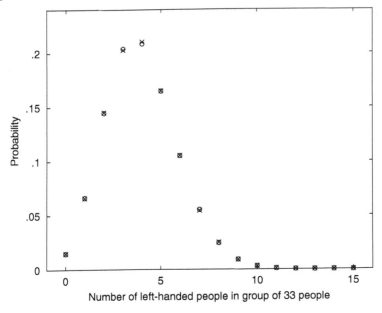

Figure 16.6 200 000 Iteration Monte Carlo simulation – 33-person group, 12% left-handed probability.

other than as to the veracity of the pseudorandom number generator. There is no dependence on normal, Student-T, Bayesian, or of any other distribution or analytic technique or information.

Superimposed on the results of the Monte Carlo simulation in Figure 16.6 are the results of a Bayesian analysis of the same problem. As may be seen, the two sets of data overlay each other so well that it's almost impossible to discern a distinction.

Figure 16.6 "looks" sufficiently normal for us to approach calculations with standard formula. Without showing the data here, from the simulation, the sample mean is 3.96 and the sample standard deviation is 1.86. Two standard deviations above the mean is therefore 7.61. Our conclusion that seeing seven people is, with a 95% confidence interval, valid.

However, is a two-sided confidence interval what we want here? Since we are only interested in the upper side of the curve, should we take the 95% point on the cumulative distribution function (CDF) rather than the 97.5% point as our criterion? Without showing the data, the 95% point is a bit less than seven people. Since we are dealing with discrete integer data, we can't come closer to the 95% point (or any other point) than the nearest integer.

Whichever path we choose, we come to the conclusion that seven people being a plausible sample of the entire population is credible, but "on the edge" of credibility. However, don't forget that 95%, 97.5%, or any other number we choose, is a choice that we feel comfortable with; it never arose from a calculation.

Now, what about the Bayesian approach to this problem? Referring back to Chapter 12 for the calculation approach, the probability of seeing seven left-handed people in a group of 33 is .055. The average of all the possible probabilities (from 0 left-handed people up to 33 left-handed people) is .029. The ratio of these two numbers is 1.9. According to Table 12.5, the result that seeing this seven-person group is OK, but "Barely Worth Mention."

The bottom line here is that both approaches come to the same conclusion.

Now let's repeat this problem, but this time we'll consider the probability of finding one left-handed person in a group of two people.

As in the previous example, a simulation and the Binomial probability calculation yield virtually identical results (not shown). Since there are only three data points (0, 1, or 2 left-handed people) for each of these functions, the CDF is a very simple graph.

We can estimate a confidence interval from this (3-point) CDF, but we have to think carefully about what our result tells us. We can say that it is most likely that we will find no left-handed people and highly unlikely that we will find two left-handed people in a group of two people. This isn't much better than a common sense estimate with no calculations whatsoever.

If we try to "do the numbers" without looking at the distribution, we return to Eq. (16.11):

$$\sigma = \sqrt{\frac{p(1-p)}{2}} = \sqrt{\frac{.12(1-.12)}{2}} = .230 \tag{16.13}$$

For a normal distribution, the upper end of a 2-sided 95% confidence interval is $(.230)(1.96) = .45$ above the mean, and for an $n = 2$ Student-T distribution it is $(.230)(12.7) = 2.92$ above the mean. However, our 3 point PDF doesn't look very normal (not shown).

Looking at the Bayesian calculation, the probability of seeing one left-handed person in a group of 2 is .211. The average of all the possible probabilities (from 0 left-handed people up to 2 left-handed people) is .222. The ratio of these two numbers is .63. Referring to Table 12.5 (looking at $1/.63 = 1.6$), we conclude that we don't think seeing one left-handed person in a group of two people is a random occurrence, but again, this is a very weak conclusion.

Correlation and Causality

Table 16.2 and a plot of the same information, Figure 16.7, show some interesting information. The table lists the ice cream sales and the robberies in some town for every month of the year. The figure plots these same ice cream sales on the x-axis versus robberies on the y-axis.

Table 16.2 Monthly data of ice cream sales and robberies.

Month	Ice cream sales ($10 000)	Robberies
January	20	76
February	26	69
March	42	121
April	49	174
May	84	209
June	68	247
July	122	287
August	93	349
September	104	342
October	94	293
November	59	243
December	31	73

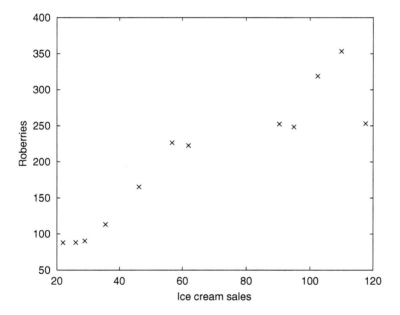

Figure 16.7 Plot of Table 16.2, monthly data of ice cream sales and robberies.

A curious observation from the table and the plot is that the ice cream sales and the robberies seem to "track" each other in some sense. When ice cream sales go up, robberies go up. When ice cream sales go down, robberies go down.

Is this propensity just a set of random happenings, or are these events somewhat *correlated*? Is there a way of quantifying correlation between two variables?

The next question, which is very important, is about *causality*: is one of these variables partially or even fully the cause of the other? If we were to plot outside light brightness versus angle of the sun in the sky, we'd be comfortable in assigning causality – when the sun is over head, it's bright out, etc. In this case, however, it's not so clear. Does eating an ice cream cone make you want to go out and commit a robbery? Or maybe it's the other way; do you celebrate having committed a robbery by eating an ice cream cone?

In this example, it's not hard to resolve the puzzle. There is no causal relationship between eating ice cream and committing robberies. The *month* column in the table gives the secret away. In warm months, people eat more ice cream than they do in cold months. In warm months, people are outside more, can move around easier, and hence commit more robberies.

The causality exists, but if the table didn't have the months column, we might never happen on its true nature. The correlation also exists: people do indeed commit more robberies in the same months that they eat more ice cream. We could say that correlation is a necessary but not sufficient condition for causality.

This example points out one of the Golden Rules of statistical analysis: correlation does NOT imply causality.

Correlation Coefficient

A measure of the correlation between two random variables is the *covariance*. For two discrete lists of the same length of random variables, X and Y, the covariance is defined as

$$\text{cov}(X,Y) \equiv EV\left\{\left[X - EV(X)\right]\left[Y - EV(Y)\right]\right\} \tag{16.14}$$

Using the same algebraic procedure as was used for the variance in Chapter 2, this is equivalently

$$\text{cov}(X,Y) = \frac{1}{n}\sum X_i Y_i - \frac{1}{n^2}\sum X_i \sum Y_i \tag{16.15}$$

If we calculate the covariance of the data of Table 16.2 using Eq. (16.15), we get 2884. This is interesting, but what do we do with it? The issue here is that this number is not scaled to anything. One way to resolve this is to use Pearson's

Correlation Coefficient, r, which normalizes the covariance to the two individual standard deviations:

$$r(X,Y) \equiv \frac{cov(X,Y)}{\sigma(X)\sigma(Y)} \tag{16.16}$$

Using the same table values again, we get $r = .91$. We still need some help to put this in perspective.

For two perfectly correlated data sets, e.g. $X = Y$ or $X = 3Y$ or $X = 3Y + 2$, we get $r = 1$. The scaling assures that r can never be greater than one.

On the other hand, for two data sets that are *anticorrelated*, e.g. $X = -Y$, we get $r = -1$. The scaling assures that r can never be less than minus one.

If two data sets are totally uncorrelated, we would get $r = 0$.

Putting all of this together, $0 \le |r| \le 1$. A magnitude of r close to zero tells us the data sets are hardly or not at all correlated. A magnitude of r close to one tells us the data sets are highly correlated – with the caveat that positive r means correlated, negative r means anticorrelated. In graphical terms, positive r means that when X goes up, Y goes up; negative r means that when X goes up, Y goes down.

Going back to the ice cream–robbery data, $r = .91$ tells us that there is indeed a strong correlation between ice cream sales and robberies. We cannot repeat too many times that this correlation tells us nothing about causality except that there might be another, hidden, variable that both the ice cream sales and the robberies are correlated to and is driving them both (is causal). In this case this hidden variable is the average monthly temperature.

The study of causality is a study which interacts significantly with statistical analysis but is a field unto itself.[3]

Regression Lines

Consider a distribution of the heights of 18-year-old men entering the army. A reasonable guess for this is a normal distribution with $\mu = 69''$, $\sigma = 2''$. If every man eventually has a son who grows to be $1''$ taller than his father, these heights would constitute a normal distribution with $\mu = 70''$, $\sigma = 2''$. Figure 16.8 shows an example of these points. The correlation coefficient is almost 1 (with a finite number of points generated by a random number generator, we never get the exact answer).

Now let's recalculate the sons' heights assuming that each son's height is derived from his father's height plus a normal distribution with $\mu = 2''$, $\sigma = .5''$ Figure 16.9

---

3 A recent introductory book is "The Book of Why" by Judea Pearl and Dana Mackenzie.

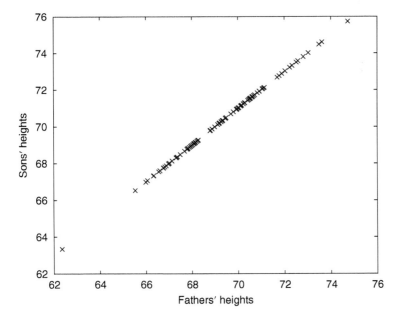

Figure 16.8 Fathers' and sons' heights distribution – ideal genetic correlation.

shows an example of these points. The correlation coefficient has dropped to approximately .95, still an excellent correlation. This figure shows points that "look like a line" but are clearly not points on a line (compare this to Figure 16.8). The line that best represents these points is called the "regression line."

The regression line is typically calculated by the "least squares error" technique. That is, the regression line is the line for which the sum of the squares of the difference between each data point and the corresponding point on the line is minimized. The formula for this is straightforward and is presented in the problems section for this chapter. Figure 16.10 shows the regression line for the data of Figure 16.9.

Regression to the Mean

Let us return to the example of fathers' and sons' heights. In the example above, each generation is, on the average, taller than the previous generation.[4] If this represented reality, we'd have some 12 ft tall people walking around by now.

---

4 This typically happens due to improved nutrition and environment, for example with immigrant populations moving to a "better life." However, for stable populations this is not the case.

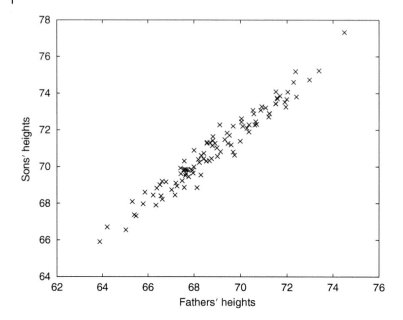

Figure 16.9 Fathers and sons heights distribution – ideal genetic + normal variation correlation.

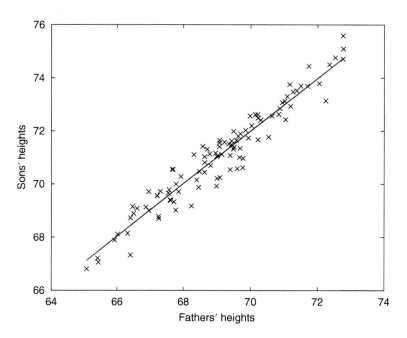

Figure 16.10 Figure 16.9 with regression line added.

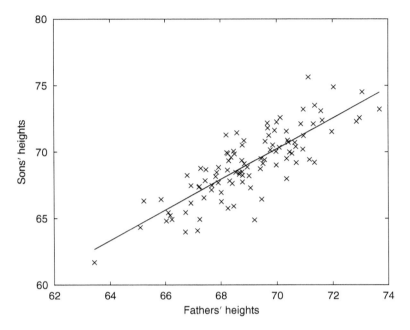

Figure 16.11 Fathers' and sons' heights – ideal genetic + normal variation predicting spreading of data over generations.

What happens if we assume that sons, statistically, have the same height as their fathers (or in general, the average of their parents' heights), but with some distribution around their fathers' heights. Figure 16.11 shows this case. The sons' heights are, on the average the same as their fathers' height, but each with his own variation about his father's height. The average height stays the same, but the standard deviation increases and the correlation decreases.[5] This means that, over generations, average height will stay the same, but we'll start seeing more and more very tall and very short people. Actually, this is not happening, the population statistics are fairly consistent over generations. Something else must be going on.

What is going on is that height is not simply an inherited characteristic. There is a component of heredity (tall parents tend to have tall children) and also a component of sheer randomness. The result is that taller sons have, on the average, sons who are still taller than average but not as tall as they are. Similarly, shorter sons have, on the average, sons who are shorter than average

---

5 It is worth repeating that these examples were generated using a (pseudo) random number generator. Unless this generator has a fixed seed, the examples will never exactly repeat – either on the author's computer or on yours.

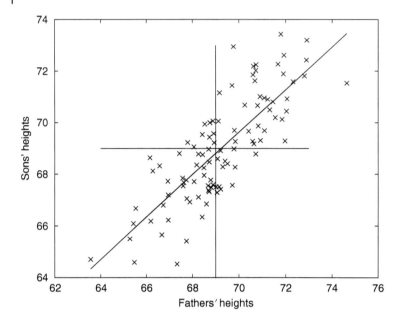

Figure 16.12 Fathers' and sons' heights – ideal genetic + random variation predicting regression to the mean.

but not as short as they are. Note that these are statistical conclusions; this does not work from a given father to his son.

Figure 16.12 shows an example of this. In the figure, the vertical line shows the sons of fathers of average height; the horizontal line shows the fathers of sons of average height. Regression to the mean works in both directions: in the previous paragraph interchange the words *fathers* and *sons* and the statements will still be true. An important conclusion here is that regression to the mean is a sheer statistical characteristic, it cannot be related to causality because a son cannot genetically be a contributor to his father's height.[6]

Problems

16.1 Consider a population of six numbers: 1, 3, 5, 6, 8, 10.

    A What is the mean and standard deviation of this population?

    B List all sample populations of size 3 that can be drawn from this population.

    C Find the means of each of these sample populations and then the mean of the sampling means. Compare this to the answer to part (a).

    D Is the result a coincidence?

---

6 The adage "Raising children can wear you down" does not have a statistical basis.

16.2 **A** Consider a population of 2000 male college students, with heights normally distributed, μ = 69.0", σ = 2.0". For 100 samples of 50 students each what would be the mean and standard deviation of the resulting sample of means?
B How many of these samples would you expect to be between 68.8 and 69.2?

16.3 You are a jelly bean manufacturer. Your jar label claims that the jar contains "At least 500 g of jelly beans." For the jelly bean jar weight population, σ = 20 g.

You would like to set up a sampling procedure for periodically carefully weighing sampled jars of jelly beans. An assumption going into this task is that you need only look at one-sided confidence intervals, no customer is going to complain about getting too many jelly beans in the jar.
A Sample 6 jars of jelly beans. What must these jars weigh for various reasonable confidences factors that say our production is under control?
B Suppose we sampled 12 jars instead of 6. What would change? (Just look at the 2σ case.)

16.4 A straight line can be represented on a graph by the equation

$$y = mx + b,$$

where x is a number on the horizontal axis, y is a number on the vertical axis, m and b are two parameters that define the line.

Consider the line

$$y = 3x + 2$$

A Plot the y points for $x = -1, 0, 1, 2, 3$, draw the line through these points.
B What are the mean and standard deviation of the x and y sets of points?
C Pretend that the y points and the x points represent data, e.g. x points are the time about midnight, y points are the temperatures in your bedroom at these times. Calculate the correlation coefficient, r, between x and y.

16.5 Calculating parameters of regression lines: these are simple, albeit messy looking, formulas. You are given two lists (the same length) of data. The independent variable data are X_i, the dependent variable data are Y_i. You want to find the parameters m and b for the regression line $y = mx + b$:

$$m = \frac{\left(\sum y_i\right)\left(\sum x_i^2\right) - \left(\sum x_i\right)\left(\sum x_i y_i\right)}{n\left(\sum x_i^2\right) - \left(\sum x_i\right)^2}, b = \frac{n\left(\sum x_i y_i\right) - \left(\sum x_i\right)\left[\sum y_i\right]}{n\left(\sum x_i^2\right) - \left(\sum x_i\right)^2}$$

where n is the length of the data lists and all sums are over $1 \le i \le n$.

Using the sample data $y1(x)$ and $y2(x)$ in the below table, plot $y1$ and $y2$ versus x. Then calculate the correlation coefficients and the parameters of the regression lines. Plot the lines over the data points on their respective graphs.

x	y1	y2
−1	−.949	17.36
0	−3.76	14.39
1	9.86	43.67
2	9.17	−64.63
3	11.38	6.41
4	13.40	−41.65
5	16.32	12.46

16.6 The below table is a repeat of the first two columns of the table in Problem 16.5 except that these columns have been interchanged ($y1(x)$ and x are reversed from that in Problem 16.5). Also, the column labels have been changed to t for the independent variable and u for the dependent variable; hopefully to avoid confusion with the original table.

t	u
−.949	−1
−3.76	0
9.86	1
9.17	2
11.38	3
13.40	4
16.32	5

Find the correlation coefficient and the regression line for this data and compare it to the corresponding results of Problem 16.5.

16.7 The below table shows your company's profits, in $million, for the past four years. These profits are all negative, i.e. your company has lost money every year. However, it does seem to be getting closer to making money each year. Based on this data, when do you think you'll start making money? How well do you believe this prediction?

Year	Prof M$
2015	−4.8
2016	3.9
2017	−3.1
2018	−2.2

16.8 You are given a coin that is purportedly a fair coin. You flip it ten times. What should your suspicions be about the actual fairness of this coin versus the number of heads or tails you flip?

17

Statistical Mechanics and Thermodynamics

Introduction

This chapter will present a very brief, nonmathematical, introduction to a field of science that is based entirely on the laws of probability and statistics. However, in most situations, the concepts of probability that we have been dealing with, such as mean, standard deviation, and confidence interval, are never mentioned. Since this science is central to how our world works, it was felt that this chapter is relevant and belongs in this book.

Our macroscopic world is made up of atomic and subatomic particles. The interactions between these particles result in what we see, hear, and touch. Gases are an important subset of these particles; gas molecules have simple individual properties (mass, velocity). However, when we are dealing with billions or more molecules of gas at once, mechanical analysis of a system is out of the question; we must deal with the statistical properties of a macroscopic volume. Two of these properties, which arise from statistical analysis, are commonly known as temperature and pressure[1]; in other words, the statistics of gases is the study of heat and heat flow, which in turn leads to the study of thermodynamics. The laws of thermodynamics teach us, among other things, how we can and cannot get useful work out of *heat engines* such as the internal combustion engine.

The air around us is made up of many, many, molecules of gas, all moving around randomly (Brownian motion). These molecules collide, and their collisions can be pictured as the collisions of hard spheres, e.g. bowling balls. Freshman physics students learn to calculate the effects of the collisions of two spheres on the trajectories and speeds of the spheres, but what can be calculated when there are billions of these spheres careening around in every

---

1 In the jargon of the philosophy of science, these are known as Emergent Properties.

Probably Not: Future Prediction Using Probability and Statistical Inference,
Second Edition. Lawrence N. Dworsky.
© 2019 John Wiley & Sons, Inc. Published 2019 by John Wiley & Sons, Inc.
Companion website: www.wiley.com/go/probablynot2e

cubic centimeter of air? The answer to this question lies in the fact that while individual gas molecules each behave quite erratically, a room full of gas molecules behaves quite predictably because when we look at the statistics of the molecules' activity, we find that the erratic behavior tends to average out. This is another way of saying that the standard deviation of, say, the x-directed velocities of the molecules involved is a very, very small fraction of the mean. The field of physics that starts with looking at the behavior of individual gas molecules and then predicts the behavior of gases is called *Statistical Mechanics* in recognition of the fact that it is concerned with building up mechanical models from the statistics of the behavior of billions of gas molecules while not being unduly concerned with detailed knowledge of the behavior of any single one of these molecules. The concepts of gas, heat, and temperature, among others, arise from this study and lead to the study of *thermodynamics*, which is concerned with the movement of heat and how we can (or sometimes cannot) get this movement of heat to do some useful work for us.

By introducing just a few principles of mechanics and using some drastic simplifications, we can derive a very good picture of the basics of gas properties and the fundamentals of thermodynamics.

Statistical Mechanics

In previous chapters we looked at many situations where the number of items involved was important to the calculations. Typically, as the number of items increased, the distributions of expected values narrowed about the mean. When we deal with the gas molecules in a room, we are dealing with so many items (gas molecules) that we don't worry about the distribution of, say, the velocity about its mean. Statistical mechanics calculations obey the same rules as do coin flips and left-handed people in a crowd, but the perspective is very different.

A gas molecule has, insofar as its macroscopic motion is concerned, only two properties: its mass and its velocity.

Velocity is, formally, speed with information about direction added in. For example, in three dimensions there will be x, y, and z directed "components" of velocity. The total magnitude of the velocity (how "fast" something is going) is called speed.

In order to reduce this discussion to its bare essentials, let's consider a simple sealed box with some gas inside. We'll put the box in outer space so that there is nothing happening on the outside. We'll ignore esoteric effects such as radiation from the walls of the box. The only things to consider are the gas molecules rattling around in the box – sometimes striking each other and sometimes ricocheting off the sides of walls of the box.

A moving mass has kinetic energy. The faster it moves, the more energy it has. It also has momentum. When some gas molecules "bounce off" the right side wall of our box, they will push the box to the right. However, at the same time, gas molecules will bounce off the left side wall of the box, pushing it to the left. The net result will be that the box doesn't go anywhere.[2] This is an important example of one of Newton's laws of mechanics, "A closed system cannot change its own momentum."

As the gas molecules rattle around, bouncing off each other and the walls of the box, they exchange momentum and energy. The total momentum, if the box is standing still, is zero.[3] The total energy is the sum of the energies of the individual particles and is fixed.

The average kinetic energy of the gas molecules is called the *temperature* of the gas. The total energy in the system is call the *heat energy* of the gas.[4] Heat energy is typically measured in Joules or Calories or BTUs. Temperature is measured in degrees Kelvin or Celsius or Fahrenheit. Degrees Kelvin (sometime called Absolute temperature) is set up so that zero degrees corresponds to zero energy. There is no such thing as a negative absolute temperature.

The force that the gas molecules push on the walls of the box with, per unit area of box wall, is called the *pressure* of the gas. If we could measure things on a small enough scale, we would see that the pressure is not a simple number; it is the average of billions of impacts every second of gas molecules with the box. Since we are dealing with so many molecules and so many impacts, as was explained above, we don't worry about this subtlety.

We can combine the above definitions with Newton's laws of mechanics and derive the Ideal Gas Law:

$$PV = nRT,$$ (17.1)

where

- P is the pressure of the gas
- V is the volume (of the container holding the gas)
- n is the amount of gas, measured in "moles.[5]"
- R is a constant, the "Ideal Gas Constant"
- T is the (absolute) temperature

---

2 If we could watch at a microscopic enough level, we would see the box "rattling" slightly from the individual molecule collisions.

3 The momentum of a particle is its mass times its velocity. Momentum, therefore, has direction components, and the momentum in, say, the $+X$ direction can cancel momentum in the $-X$ direction. This results in 0 net momentum for the system even though individual particles have momentum.

4 The English language muddies these concepts. The word "hotter" means "is at a higher temperature." It does not mean "has more heat (energy)."

5 One mole of particles of any substance contains 6.022×10^{23} particles.

We have now totally left the realm of statistics and moved into the realm of chemistry.

In a solid, the picture is a bit more complicated than it is in a gas, although the concepts remain the same. A solid is a collection of atoms that are bound together by chemical bonds, aka interatomic forces. The atoms of a solid are not free to move as they are in a gas.

If we picture the interatomic binding forces as little springs,[6] then we can imagine atoms in a solid with some kinetic energy vibrating in place, periodically exchanging their kinetic energy with potential energy stored in the compression or extension of these springs. Temperature is now the average of the sum of the kinetic and potential energies.

(Concepts of) Thermodynamics

What happens when we bring two collections of atoms together – either by combining two gases in a common container or by pressing two solids up against each other? In a gas, the different molecules just fly around and interact with all the other particles and the container wall. In a solid, the vibrating atoms in each body transfer vibrational energy to the other body.

In both situations, the total energy of the system is the sum of the individual energies. The temperature, aka the average energy, must be somewhere between the original higher temperature and the original lower temperature. If we wait a while, the new combined system reaches a uniform temperature and the system is said to be in "thermal equilibrium."

Looking at the solid, when the two solids are first brought into contact with each other, both solids transfer vibrational energy across the interface to the other solid. The net flow of vibrational energy must be from the system that had a higher average vibrational energy, i.e. that was originally at the higher temperature of the two. After a while, the average energies are the same and even though vibrational transfer continues to occur, the average transfer at the interface is zero – the two competing transfers cancel out. This is another way of saying that both bodies are now at the same temperature.

Looking at the gas, we reach the same conclusion, although the energy transfers are occurring at individual atomic collisions.

We can generalize the above into a very basic concept of heat flow: when two bodies at different temperatures are allowed to exchange energy (heat), they will do so. Net energy flow will be from the hotter body to the colder body until both bodies are in thermal equilibrium (at the same temperature).

The characteristics described above allow us to design a machine that will convert heat energy into mechanical energy, called a *heat engine*.

---

6 This picture is insufficient for mechanical calculations such as of elasticity, but it is useful as a conceptual image.

Figure 17.1 is a sketch of a simple heat engine. Energy is put into the system (the fire) which heats the gas in the container. Since the volume and amount of gas, at the outset, is fixed, according to the ideal gas law, the pressure of the heated gas must go up. Gas at a higher than ambient pressure pushes out of the small hole in the container and spins the paddlewheel (shown in cross-section). The paddle wheel shaft might run a locomotive, or be connected to an electric generator.

When the gas pushes the paddlewheel, energy is transferred to the paddle-wheel, cooling the gas. Note that room-temperature-and-pressure air are pushing at the back side of the paddlewheel blade. Some energy is expended pushing this air out of the way, so the energy put into the system is never fully converted to useful work. Since the net force to turn the wheel is proportional

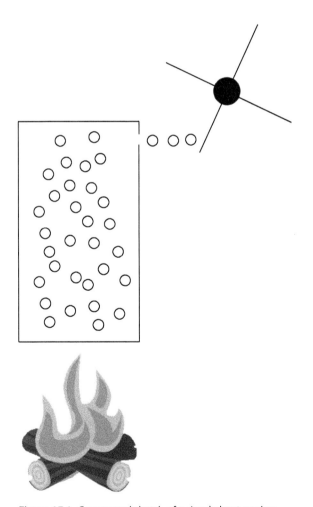

Figure 17.1 Conceptual sketch of a simple heat engine.

to the pressure differential which in turn is proportional to the temperature difference between the heated and the room-temperature gas, the efficiency of the system is proportional to the temperature difference between the heated and the ambient air.

We have just seen two basic properties of heat engines: you can never get all the energy in the system to do work for you unless you have an absolute zero temperature low-temperature region to cool things. And, given that the low temperature side is fixed by the ambient temperature, the higher you can heat the high temperature side, the more efficient your heat engine will be.

Most heat engines work using periodic cycles of heating a gas, having the gas push on and move something, expelling the cooled gas, and then repeating. The details of these cycles typically bear the inventor's name: the Carnot cycle (aka the internal combustion engine), the Diesel cycle, the Wankel cycle, etc.

There is another basic thermodynamic system property which is exemplified by the act of pouring some hot water into a partially filled glass of cold water. No work has been done. We now have a glass of warm water whose total energy is equal to the sum of the energies of the partial glasses of hot and cold water.

Now let's put it back the way it was; we want to separate the glass of warm water into two partial glasses of water, one hot and the other cold. Even though no energy was gained or lost when we mixed the water, we will need to supply energy to separate the water back to the way it was originally.

In other words, the system is inherently irreversible. The irreversibility of a thermodynamic process is characterized by its *entropy*. When a system evolves toward thermodynamic equilibrium, its entropy increases. Thermodynamic equilibrium is the state of maximum entropy.

Entropy can also be defined in terms of the order/disorder of a system. As disorder increases, entropy increases. These two definitions are equivalent, they are expressing the same physical property in terms of macroscopic (heat) terms or microscopic (statistical) terms.

If we mix two partially filled glasses of water that are originally at the same temperature, then we do not change the entropy. We can separate the water back into the two glasses without adding any energy (doing any work). This system is reversible.

We now have another principle of thermodynamics: a closed system will naturally evolve to its highest entropy state. To lower the entropy of a closed system, we must expend energy. A good example of this latter situation is the inside of a refrigerator. It is in a lower entropy state than the room that the refrigerator is sitting in. Every time we open the refrigerator door, the outside and inside gases mix and start evolving toward equilibrium. We must close the refrigerator door and expend some energy to cool the inside of the refrigerator again.

A refrigerator cools its inside by expelling heat to the outside room. If the same system is assembled to cool a room by expelling heat to the outside environment, we call it an *air conditioner*. If we turn it around and transfer heat from the outside environment into a room, we call it a *heat pump*.

Since any system where heat can be exchanged between its "particles" will evolve toward its highest entropy state, at which time everything will be at the same temperature and no more work can be done, what about our universe? Will there not be an ultimate "heat death of the universe?" This discussion is outside the scope of this book.

The discussions above are summarized by the three laws of thermodynamics, which have both a formal and a "common" form. Formally:

1) Conservation of Energy: the total energy in the universe is constant. Of more practical value, in a closed system, that is a system in which energy cannot enter or leave, the total energy must be constant.
2) The entropy of any (closed) system that is not in equilibrium (not at a uniform temperature) will increase with time. In the simple heat engine example above, if we generate the high pressure steam and then just let it dissipate into the atmosphere, we haven't lost any of the heat energy that was put into the steam. Instead, we've cooled the steam a lot, heated the local air a little, and totally lost the ability to get any useful work out of the steam. So long as the hot steam and the cool ambient air are separated, we have an ordered system with a low entropy and an opportunity to get some useful work done. Once the hot steam and the cool ambient air are mixed together, we have a disordered system with a higher entropy, and have lost the ability to get useful work done.
3) We define an absolute zero of temperature as the temperature at which all thermal motion ceases. In general, cooling a system takes work and both the heat resulting from this work and the heat removed from the system must be "dumped" somewhere else. Cooling a region lowers the entropy of the cooled region even though the entropy of the universe is increasing because of the effort. If a region could be cooled to absolute zero, its entropy would be zero. The 3rd law says that it is impossible to reduce any region to absolute zero temperature in a finite number of operations.

Most students of this material lose familiarity with these laws over the years and could not easily quote them to you. However, virtually all of them can quote a simple paraphrasing that is surprisingly accurate:

> First Law of Thermodynamics: You can't win (i.e. you can't create "new" energy)
>   Second Law of Thermodynamics: You can't break even (i.e. you can't run any heat machine without wasting some of your energy)
>   Third Law of Thermodynamics: You can't get out of the game (i.e. you can never reach the zero entropy state of absolute zero temperature)

18

Chaos and Quanta

Introduction

Chaos theory is a branch of mathematics which considers, among other things, how very erratic and unpredictable looking results can arise, under certain conditions, from very simple systems. The "If a butterfly flaps its wings in Brazil, it can set off a Tornado in Texas." tale is a popular story about this phenomenon. Chaotic systems may look as though they are based on random phenomena when in fact they are completely deterministic. This topic is being presented here as an example of something that looks random when it fact it is not.

Quantum mechanical phenomena, which we see at microscopic scale, ultimately determines all physical phenomena. As is best understood today, the basic properties of subatomic particles, e.g. the position of an electron, are determined by probabilistic distributions. The random events that these distributions are describing are truly random – not pseudorandom – events. As possibly the only truly random phenomena in the universe, how could they not be included in a book about probability?

Chaos

Figure 18.1a is the output of a (pseudo)random number generator; we are looking at 100 numbers between 0 and 1. Compare this to Figure 18.1b which is 100 values of the (recursive)[1] function

$$x_{new} = ax_{old}\left(1 - x_{old}\right) \tag{18.1}$$

---

1 "Recursively" means you feed the output of a function back into the input and get a new output and then do this again, and again....

Probably Not: Future Prediction Using Probability and Statistical Inference,
Second Edition. Lawrence N. Dworsky.
© 2019 John Wiley & Sons, Inc. Published 2019 by John Wiley & Sons, Inc.
Companion website: www.wiley.com/go/probablynot2e

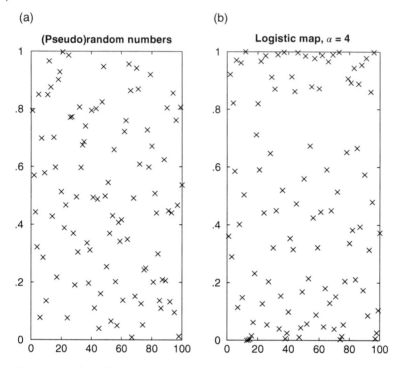

Figure 18.1 Plots of 100 pseudorandom numbers and a logistic map set, 100 numbers.

For this figure $\alpha = 4$, x_{old} was seeded at .1, and the formula was run 500 times before the 100 values to be graphed were recorded. This function is known as a "Logistic Map" generator.

Both of these figures look like lists of random data. If they weren't labeled, it would be impossible to tell by looking which is which. Both of the lists have the correct mean and sigma values for a set of random numbers between 0 and 1. Is there a difference between them or is Eq. (18.1) an easy way to generate pseudorandom numbers? The generated list of numbers doesn't seem to show any simple repetition patterns, or anything else obvious.

Figure 18.2 uses the same lists of numbers used in Figure 18.1, but plotted in what is called Phase Space. These graphs are generated by taking each number on its list (except the last) and using it as the X, or horizontal axis, value and then taking the next number on the list and using it as the Y, or vertical axis, value. Table 18.1 shows how this works for the first 11 numbers on the list.

The graph of pseudorandom numbers (Figure 18.2a) doesn't look different from that in Figure 18.1a; everything still looks random. However, Figure 18.2b is clearly different from Figure 18.1b – it does not look like a plot of random combinations of numbers. We have created a situation in which deterministic data looks random, unless it is looked at carefully.

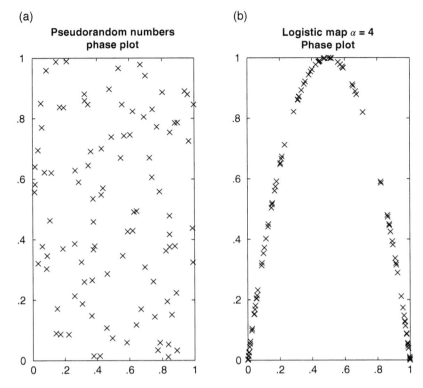

Figure 18.2 Phase space plots of 100 pseudorandom numbers and a logistic map set, 100 numbers.

Table 18.1 Creation of phase space plot from a data list.

Data	x	y
		Plotted points
.4504	.4504	.9902
.9902	.9902	.0389
.0389	.0389	.1497
.1497	.1497	.5091
.5091	.5091	.9997
.9997	.9997	.0013
.0013	.0013	.0053
.0053	.0053	.0212
.0212	.0212	.0829
.0829	.0829	.3041
.3041		

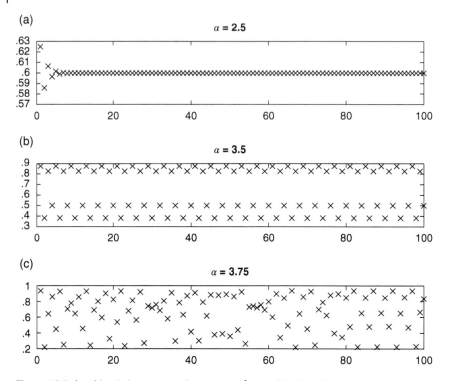

Figure 18.3 (a–c) Logistic mat equation outputs for α = 2.5, 3.5, 3.75.

The above introduction shows that the output of the logistic map generator function is unusual. However, we have not yet shown how this relates to a discussion of chaos.

Return to the *logistic map* equation above. Figure 18.3a–c shows what the (recursive) logistic map equation produces for 50 recursions, using 3 different values of the parameter α. In each case we started, or *seeded*, the equation with the value x = .4. For α = 2.5 (Figure 18.3), x quickly settles to a fixed value of .6. This is not exciting. For α = 3.5 (Figure 18.3b), x never settles to a fixed value, but rather takes on what is call *periodic* behavior. That is, it repeatedly cycles between (in this case 4) fixed values.

At α = 3.75, all hell breaks loose (Figure 18.3c). This is the type of behavior shown earlier that looks random until you draw a phase plot.

The behavior of a function such as the logistic map can be summarized in what is called a *bifurcation diagram*, shown in Figure 18.4.

This diagram was generated by taking successive values of the parameter a; for each value of a, run the recursive function (the logistic map) 50 times just to make sure that the function would settle down if they were going to settle

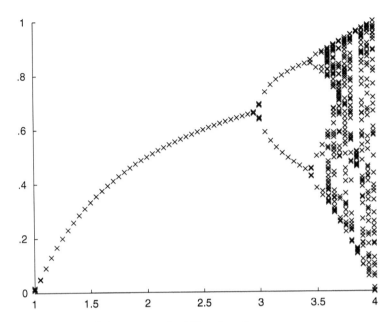

Figure 18.4 Logistic map equation bifurcation diagram.

down, and then running the recursion function 50 times more and plotting the resulting values of x.

For α less than about 3, the 50 values of x (at each value of α) are identical, the points lie right on top of each other. For α between about 3 and 3.4 there is a simple periodicity – the values of x bounce back and forth between a high value and a low value. For α = 3.5 we see the slightly more complex periodicity described above – the values of x bounce back and forth between 4 different values. Then, finally, for α greater than about 3.7, we see the truly chaotic behavior of this function.

Now we can return to the question of why we care.

Figure 18.4 illustrates the "Butterfly Wing" effect. A very slight change in α causes drastic changes in both the number of values of x and the values themselves. Now, what if a much more complex equation or system of equations had similar behavior. Let's say that α is related to the average temperature and x to the peak winds of hurricanes around the world in a given year.[2] If α is small enough, there are no hurricanes that year. If α is a bit bigger, there will be 1, or 2 storms. If there are 2 storms, 1 of them will be moderately strong. If α is a bit

---

2 This is certainly NOT real meteorology. It's just a hypothetical example to demonstrate the possibilities and implications of chaotic behavior.

bigger yet, there will be an unpredictable number of storms with wildly varying intensities. Furthermore, in this latter chaotic region, the subtlest change in α can cause drastic changes in both the number and intensities of the storms.

Another implication of chaotic behavior is referred to as *fractal behavior*.

Suppose we have good weather predictions for tomorrow's high temperature in San Francisco, California, and New York City, New York, and we want a prediction for the high temperature in Des Moines, Iowa. Even though Des Moines is approximately half-way between San Francisco and New York City, it makes very little sense to say that the Des Moines high temperature will be about the average of the New York and San Francisco (high) temperatures.

On the other hand, suppose we had weather stations about 10 miles to the west of Des Moines and 10 miles to the east of Des Moines. It would be reasonable then to predict Des Moines high temperature (approximately) as the average of these two weather stations' predictions.

This approach, called *interpolation*, is shown in Figure 18.5. In Figure 18.5a, suppose we knew the values of our function at points a and e and we wanted an approximation of the value of the function at point c. Taking the average of the values at a and e would work well.

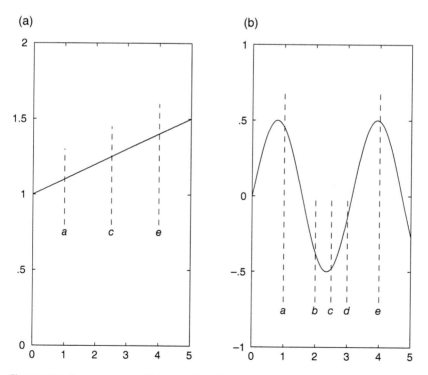

Figure 18.5 Two examples of interpolation of measurements.

Figure 18.5b shows a function that is not simply a straight line, as was in Figure 18.5a. In this situation, averaging the values at points *a* and *e* would give us total nonsense. Averaging the values at points *b* and *d*, which are much closer to point *c* than are points *a* and *e* would not give us an accurate answer, but it would give us a reasonable approximation. Bringing points *b* and *d* closer together about point *c* would improve the accuracy of the approximation.

This whole approach falls apart when we have a chaotic behavior known as *fractal*. Figures 18.6 are of the *Koch Snowflake*. The Koch Snowflake is built on a recursive procedure. We start with an equilateral triangle ($n = 1$). We divide each leg into three equal segments, we build a new equilateral triangle on the center of these segments, then we remove this center segment ($n = 2$). Then we repeat this procedure as many times as we'd like ($n = 3, 4, \ldots$).

The Koch Snowflake has very interesting topological properties. As *n* gets larger, the area approaches a constant value while the perimeter keeps growing without a limit.

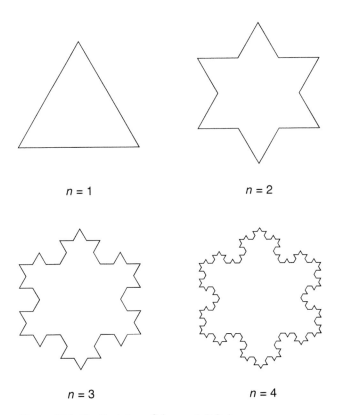

$n = 1$ $n = 2$

$n = 3$ $n = 4$

Figure 18.6 The Koch Snowflake, $n = 1, 2, 3, 4$.

(a)   (b)

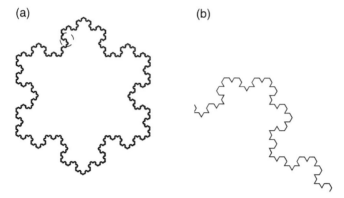

Figure 18.7 The $n = 7$ Koch Snowflake.

Figure 18.7a shows the $n = 7$ Koch Snowflake. At this magnification it doesn't look different than the $n = 4$ snowflake shown in Figure 18.6. Figure 18.7b shows a 25× magnification of the small region in the circle near the top in Figure 18.7a. The point here is that the structure keeps repeating at a finer and finer scale. In principle, we could draw an $n = 1000$ snowflake. It would look no different to the eye than the $n = 4$ snowflake. However, we could zoom in 10, 50, and 100 times and each magnification would look exactly the same to the eye. We lose the ability to discern scale with fractal surfaces. Also, we lose the ability to estimate perimeter with our eye; while the $n = 4$ and $n = 100$ snowflakes look the same, they have greatly different perimeters. The actual perimeter is called the Fractal Length. For example, if we start ($n = 1$) with a triangle of perimeter 1, the $n = 100$ snowflake will have a perimeter of almost 10^{13}! The area of the $n = 100$ triangle, on the other hand, is less than twice the area of the $n = 1$ triangle and never gets much larger.

The inability of our eyes (and brains) to discern scale when looking at a fractal figure is most often seen when gazing at mountain ranges in the distance. Figure 18.8 is a simulation of this effect. We are looking at a mountain range at three different magnifications, i.e. "zooming in" from magnification 1 to 5 and then to 25 (note the numbers on the x axis). Each magnification reveals a granularity that was not visible at a lower magnification but which, to the eye, looks the same.

This effect explains why it is difficult to tell whether you are looking at a large mountain range far away or a smaller mountain range up close. Sometimes seeing a car at the foot of the range calibrates your eye and creates perspective because you know how big a car is. Otherwise the natural fractal nature of many mountain ranges makes it almost impossible for your brain to establish a scale.

The discussion above, above chaos and fractals, was presented as a contrast to the random or pseudorandom phenomena that the rest of this book is about.

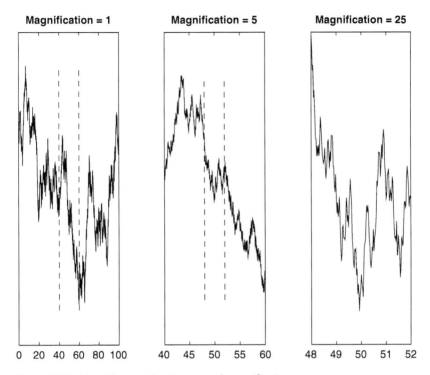

Figure 18.8 A fractal mountain view, several magnifications.

It was included because the "random look" of many of these phenomena makes it easy to be fooled. There is nothing random in anything that was discussed.

Online websites showing the Mandelbrot Set present a beautiful example of fractal repeating behavior.

Probability in Quantum Mechanics

Some atoms exhibit *radioactivity*. This means that they emit particles (electrons or helium nuclei) or electromagnetic radiation (gamma rays) and the atoms change to another type of atom. This process is known as *radioactive decay*. As an example, assume we have a material whose atoms each has a probability of .01 (1%) of radioactive emission per day. Let's start with 10^{23} of atoms that emit electrons. This might seem like a very large number (compared to most of the things we actually count every day, it **is** a very large number), but it is approximately the number of atoms in a sugar-cube sized lump of material.

Since we're dealing with such a large number of atoms, we can treat the expected value of atoms decaying as exactly what's going to happen. On the

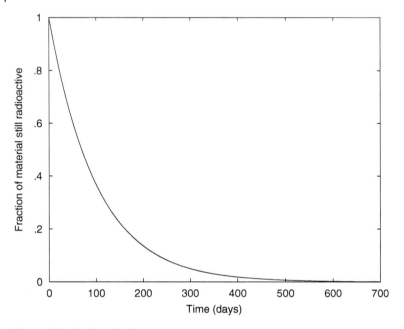

Figure 18.9 Radioactive decay.

first day this material would emit $(.01)(1 \times 10^{23}) = 1 \times 10^{21} = .01 \times 10^{23}$ electrons and this same number of particles would no longer be radioactive. There would then be $1 \times 10^{23} - .01 \times 10^{23} = .99 \times 10^{23}$ atoms of radioactive material left.

On day two, the material would emit $(.01)(.99 \times 10^{23}) = .0099 \times 10^{23}$ electrons and there would be $.99 \times 10^{23} - .0099 \times 10^{23} = .9801 \times 10^{23}$ atoms of radioactive material left. Figure 18.9 shows the fraction of radioactive material left as a function of time for about 2 years. Mathematically, this figure is the curve[3]

$$y = e^{-.01x} \tag{18.2}$$

Note that, although we are only dealing with a small decay probability each day for any given atom, we are dealing with many atoms at once. The standard deviation of the number of atoms decaying in, say, the first day is

$$\sigma = \sqrt{np(1-p)} \approx \sqrt{np} = \sqrt{\mu} = \sqrt{1 \times 10^{23}} \approx 3 \times 10^{11} \tag{18.3}$$

This seems like a very large number, but as a fraction of the mean (np), it is $3 \times 10^{11}/1 \times 10^{23} \approx 3 \times 10^{-12}$, which is a very small number. This is why, even though the "performance" of an individual atom is very erratic, the observable decay rate of even a small lump of material is a very smooth curve.

---

3 This function is known as *exponential decay*, discussed in the appendix. It describes the situation where the rate that something is decaying is directly proportional to the amount present.

After approximately 69 days, half of the starting material is still radioactive. This material is therefore said to have a "69-day half-life."[4] Actual material half-lives vary wildly. For example, one form of astatine has a half-life of about .03 seconds while one form of beryllium has a half-life of over a million years!

Having characterized radioactive decay, physicists then started asking what prompts an individual atom to decay? The amazing answer, couched in very un-scientific terms, is that "it decays when it wants to." That is, this very basic characteristic of some types of atoms is a truly random process. It is governed by the laws of probability as a very fundamental fact of nature. We have here a phenomenon that doesn't just *look* random such as the output of a pseudorandom number generator, but really *is* random, with a PDF for radioactive decay that is as much a part of the fundamental nature of this atom as the number of electrons orbiting the nucleus.[5]

The concept that random events are at the very core of the mechanics of the universe was not accepted readily as it evolved in the early twentieth century. Albert Einstein, one of the key contributors to the early quantum theory, purportedly never accepted it and left us with the famous quote that "God does not play dice."

One of the paths investigated to look for determinism rather than randomness was the idea of *hidden variables.* This is directly analogous to the operation of a pseudorandom number generator, but looking backward: a good pseudorandom number generator will produce a sequence of numbers that pass all your tests for randomness until someone shows you the algorithm that produced the numbers. Then you realize that you have been dealing with a totally deterministic sequence of numbers that just looked random. Physicists pursuing this path hoped to extend their understanding of atomic physics to a deeper level that would reveal some deterministic rules that just happened to produce results that looked random. They have never (at least never yet) succeeded.

Another attribute of quantum mechanics that is very observable in the right circumstances is electron tunneling: in a metal, electrons of sufficient energy can move around freely. This is, in a sense, analogous to a bouncing ball rattling around in a walled yard. There is an energy barrier that keeps the electrons from leaving the metal – again analogous to the ball rattling around in a walled yard. If the metal is heated, some electrons gain enough energy to overcome the energy barrier and escape the metal (some balls bounce over the wall). This is called thermionic emission and is the basis of electronic vacuum (radio) tube operation.

Quantum mechanically, the position of the electron in a metal can be described as a wave. The mathematical description of this wave is the solution

---

4 The actual relationship is *half life* $= Ln(2)/p \sim .693/p$ where p = rate of decay per unit time.
5 It's not hard to picture how a radioactivity counter and a small piece of radioactive material could be combined to form a true random number generator.

to the equation developed by Erwin Schröedinger and which carries his name. The physical interpretation of this wave is that the probability of the location of the electron is monotonically related to the amplitude of this wave.[6] Once again, we see probability as a real physical property.

The wave function of an electron in a metal does not end abruptly at the surface of the metal. The wave function penetrates the energy barrier at the surface, falling off very rapidly until it is almost zero outside the metal. This means that there is a nonzero probability that the electron can get outside the metal. Normally this probability is so small that it is ignorable – electrons do not escape from the metal.

Applying an electric field to the metal narrows the surface energy barrier. For barriers thin enough, the probability that the electron will show up outside the box grows and electrons show up outside the metal.

Classical (Newtonian) physics cannot explain this phenomenon. The wave function and related probability describe something that has survived innumerable experimental validations. One of the first successes of quantum theory to predict something was the Fowler–Nordheim equation. This equation predicts how many electrons (i.e. what electric current) will flow from a metal surface into a vacuum as a function of the height of the potential barrier at the surface of the metal and the strength of an applied electric field. Fowler and Nordheim showed that an applied electric field distorts the shape of the potential barrier and makes it thinner. They correctly predicted the electric current as a function of the applied field.

Electron tunneling is used today in scanning electron microscopes. A small, sharp, metal tip (at which an electric field concentrates) is used as the source of the electron stream that illuminates the object being studied. Electron tunneling in semiconductors is used in *flash memories*. These are the computer memories that are used in the small "USB Dongles" that we plug into our computers to save files of pictures or whatever. In a flash memory, a voltage is used to narrow a potential barrier so that electrons can transfer onto an isolated electrode. When this voltage is removed the potential barrier widens again, the tunneling probability drops very, very low and the electrons are trapped on this electrode – hence the memory effect. There are of course millions or billions of these isolated electrodes on one small chip, giving us our "Mbyte" and "Gbyte" memories. Associated circuitry senses the electron level on trapped electrodes and "reads" the memory without disturbing its content.

Discussions of the philosophical implications of quantum mechanics are fascinating. It's quite possible that truly random events and the laws of probability run the universe.

---

6 The wave is a complex number, carrying magnitude and phase information. The probability of the wave being in a given position is $(A^*)A$ where A is the amplitude of the wave and A^* is its complex conjugate.

Appendix

Introduction

In order to "tell our story" adequately, there are several instances in the text where it is necessary to use some topics from calculus. For people who have not studied calculus, this is bad news.

The good news is, in the limited context that these calculations are needed, it is only necessary to understand what is being calculated, not how to do the calculations.

The purpose of this appendix is to introduce two concepts from calculus and explain, in the limited context of this book, what is being calculated. This is in no way a substitute for learning calculus, but it will be enough for a reader to understand what is being calculated and why the calculation is necessary. In order to actually perform all of these calculations, it's necessary to study calculus.

Continuous Distributions and Integrals

When we are dealing with a finite discrete probability distribution function (PDF), there is a list of values, x_i, and their associated probabilities, p_i. Since the list of values includes every possible occurrence (e.g. the integers from 2 to 12 for the rolling of a pair of dice), we know that since one of these occurrences must occur, the sum of all the probabilities must = 1, the *certain event*. In mathematical notation, we say that

$$1 = \sum_i p_i$$

If we have a list of probabilities that does not add up to 1, we scale, or normalize, all of these probabilities so that they do add up to 1.

Probably Not: Future Prediction Using Probability and Statistical Inference,
Second Edition. Lawrence N. Dworsky.
© 2019 John Wiley & Sons, Inc. Published 2019 by John Wiley & Sons, Inc.
Companion website: www.wiley.com/go/probablynot2e

When we have a continuous PDF, such as the probability of the temperature outside your front door at noon today, then we must handle things differently. We now have an infinite list of possible occurrence values (20 degrees, 30.2, 42.67, 42.678, etc.) and we cannot talk about the probability of a given temperature – even to specify a "given temperature" would require an infinite number of decimal places in our specification!

Figure A.1 shows a possible temperature distribution function.

The curve in Figure A.1 looks normal (Gaussian), but isn't. It is only non-zero for $-10 \leq T \leq 110$. In its nonzero range, it is a normal curve minus the normal curve value at $T = -10$ (also at $T = 110$). This forces the curve's values at $T = -10$ and $T = 110$ to be 0. This was done to avoid having a curve that goes out in both directions to infinity and forcing us into a debate as to what the temperature in these "far out" regions might mean. Two temperatures, $T = a$ and $T = b$, are, labeled in the figure for reference below.

For continuous functions such as the function in Figure A.1, we will not try to talk about the probability of a Specific Temperature, but instead talk about the probability of the Temperature between $T = a$ and $T = b$. a and b can be any temperatures. This probability is proportional to the area under the curve (Figure A.1) between these points. We'll require that $a < b$ just to avoid having to untangle negative temperature range issues.

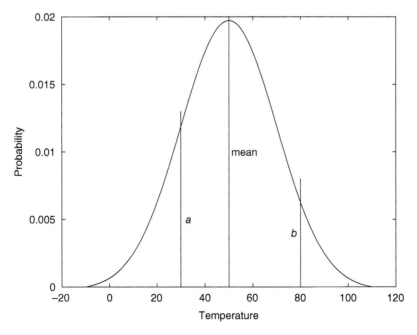

Figure A.1 Example of a finite width, continuous distribution function.

Describe the area under the curve of a function $f(x)$ over the range $a \le x \le b$ as "the integral from $x = a$ to $x = b$ of $f(x)$" and write it as

$$\int_a^b f(x)dx$$

"dx" in the above equation specifies that the lower and upper limits of integration are $x = a$ and $x = b$, since there might be other variables inside $f(x)$ (in the jargon, we are *integrating over x*).

Since $f(x)$ is a PDF, the area under the entire curve must be 1 (again, the *certain event*). If we put a scaling, or normalizing, number in front of $f(x)$, then we may write

$$1 = A \int_{-\infty}^{\infty} f(x)dx$$

And if we can evaluate the integral, we can find A so that $Af(x)$ is indeed a PDF.

Note that we said IF we can evaluate the integral. This is not a given. There are handbooks of evaluated integrals of different functions that have been built up over the years as people figured them out. There are some functions that do not have "closed form" evaluations. Integrals that cannot be evaluated in closed form can always be approximated numerically, so we do not have practical issues with evaluation.

Learning how to evaluate integrals is beyond the scope of this little tutorial. However, there are certain calculations that we can do. Assume that Figure A.1 is already normalized, i.e. $A = 1$. Then,

$$0.5 = \int_{30}^{mean} f(x)dx = \int_{mean}^{80} f(x)dx$$

because this function is symmetric.

$$0 = \int_{-\infty}^{30} f(x)dx$$

Because this function is 0 for $x \le 30$

If we define a new function $F(u)$ as

$$F(u) = \int_{-\infty}^{u} f(x)dx$$

then

$$F(+\infty) = \int_{-\infty}^{+\infty} f(x)dx = 1$$

$F(u)$ is identically the cumulative distribution function (CDF) associated with the PDF $f(x)$. This means that, for a normal distribution, we can simply "look up" any integral of a normal distribution. This CDF is usually referred to as the "error function," errf(z). We have already used this property for standardized normal distribution confidence interval calculations. For example, suppose we want the integral from 2 to 5 of the standardized normal distribution $f(z)$:

$$\int_2^5 f(z)\,dz = \int_{-\infty}^5 f(z)\,dz - \int_{-\infty}^2 f(z)\,dz = F(5) - F(2)$$

Another thing to remember is that for many distributions, e.g. a uniform or a triangular distribution over some range of x, integrals, i.e. areas under the curve, may be found geometrically – formal calculus calculations are not necessary.

Exponential Functions

The function

$$f(x) = 2^x$$

is a monotonically increasing function of x. Figure A.2a shows this function (solid line) and its rate of change (dashed line). The rate of change tracks the function closely, but is always less than the function itself.

Figure A.2b shows the function

$$f(x) = 3^x$$

and its rate of change. Again, the rate of change tracks the function closely, but here is always greater than the function itself.

These two figures imply that somewhere between 2 and 3 is a number where the rate of change is, at least approximately, the same as the function.

Figure A.2c shows just this function. The number we are looking for comes up often enough that it is given a name, "e," and the function is

$$f(x) = e^x$$

where $e \approx 2.718$.[1]

In Figure A.2c you cannot see the dashed line showing the rate of change of e. That is because this rate of change is not just approximately, but is exactly, the same as the function e itself. This property is a defining characteristic of e.

---

1 e is a transcendental number. It cannot be expressed exactly with a finite number of digits (π is another example of such a number).

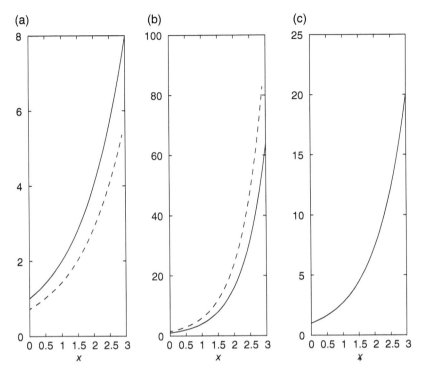

Figure A.2 The function a^x (solid line) and its rate of change (dashed line) for the cases $a = 2, a = 3, a = e$.

In general, for all three of the examples above and for any (positive) number raised to an integer, the rate of change of the function will be equal to the function times a scaling factor. For e, this factor is 1. For a number smaller than e, the factor is between 0 and 1; for a number bigger than e, the factor is greater than 1. These functions are said to exhibit "exponential growth." If the exponent is negative, these functions are falling (toward 0) functions of x and exhibit "exponential decay."

There are many other interesting properties that e exhibits that are not directly related to the discussions in this book, they will not be presented here.

Index

Probably Not: Future Prediction Using Probability and Statistical Inference,
Second Edition. Lawrence N. Dworsky.
© 2019 John Wiley & Sons, Inc. Published 2019 by John Wiley & Sons, Inc.
Companion website: www.wiley.com/go/probablynot2e